WIND POWER

WIND POWER

and Other Energy Options

David Rittenhouse Inglis

Ann Arbor
The University of Michigan Press

Copyright © by The University of Michigan 1978
All rights reserved
Published in the United States of America by
The University of Michigan Press and simultaneously
in Rexdale, Canada, by John Wiley & Sons Canada, Limited
Manufactured in the United States of America

1982   1981   1980   5   4   3

Line illustrations prepared by Rosa Patino and William Graham

Library of Congress Cataloging in Publication Data

Inglis, David Rittenhouse, 1905-
    Wind power and other energy options.

    Bibliography: p.
    Includes index.
    1. Wind power.   2. Atomic power-plants.
3. Power resources.   I. Title.
TK1541.154   621.4   78-9102
ISBN 0-472-09303-7
ISBN 0-472-06303-0 pbk.

Preface

Most of recorded history has been shaped by the use of wind power for the propulsion of ships. Furthermore, the early part of the industrial revolution made extensive use of wind power for industrial purposes. The extremely rapid industrial growth of the past century or two, however, has been based on the profligate exploitation of the fossil-fuel heritage from eons past—a nonrenewable, one-time splurge for mankind. Applications of science provided new technologies, such as electric power transmission to add to the convenience and magnitude of the splurge. Viewing the impending end of the fuel supply four or five decades ago, one wondered if science would provide a way to continue.

One possible technical answer to the shortage of fuel would be a rapid introduction of solar-related power sources, starting with one that requires no further research and is ready to be utilized now, that is, wind power. Electricity was generated by wind power as early as 1893 in Denmark, and early in this century, wind provided an appreciable part of the commercial electric power there. During the intervening years, when fossil fuels have been plentiful and cheap, a few forward-looking individuals and organizations in various countries have investigated and developed the possibility of harnessing the wind to generate electric power. It is a renewable resource with no apparent limit, for the wind will continue to blow.

Modern engineering gradually changed the form of the large wind dynamos, making them differ drastically from the familiar windmills of earlier centuries. The largest unit demonstrated was in Vermont during World War II, just before the advent of the nuclear age, and others almost as large were developed in France a decade or two later. These units proved that wind generation of commercial electric power was technically practical, but because fossil fuel was cheap and nuclear power loomed for the future, they seemed economically unattractive.

Indeed, two or three decades ago nuclear power seemed to be science's answer to the threat of exhausting fuel supplies, and at first, there were euphoric hopes that it would provide cheap, trouble-free, and almost limitless power. Events since then have raised grave doubts.

It is difficult to start large enterprises quickly even when they seem wise, and it is more difficult to stop them when they seem unwise. The commercial use of nuclear power was difficult to initiate. Having developed the technology of nuclear power plants in national laboratories as a spin-off from wartime military activities, the government had to offer substantial inducements to private industry in order for the private sector to take over the financing and construction of the nuclear power plants. At the same time, the government had to continue to fund important steps in the nuclear fuel cycle in government plants at the taxpayers' expense. Even with continuing subsidies, the industry is now in bad shape and is calling for more financial aid. A large part of the motivation for the earlier government promotion was the military demand for plutonium to be produced in commercial power plants. Since this demand has been more than satisfied and the government stopped buying back plutonium over a decade ago, the taxpayer subsidy now only helps make nuclear power seem less expensive than it actually is.

As the number of commercial nuclear power plants has grown, their cost has been rapidly increasing and they have been plagued with many more minor malfunctions than were anticipated. There have even been some grotesque instances of nuclear power plants going wildly out of control. Their radioactivity burden is potentially so hazardous that it must be contained with a high degree of confidence; yet the rate of failure raises serious doubts about reactor safety. These doubts have been glossed over by official studies that appear to have been biased and selectively reported. This adds to the difficulty of judging the extent of the danger. Along with the realization that the safety of a nuclear power plant cannot be judged by the large amount of time and money that has been devoted to it, there is a growing appreciation of the fact that the greater availability of nuclear materials accompanying the growth of nuclear power makes nuclear terrorism and even nuclear war more likely, in spite of attempts to develop safeguards.

With its rapidly increasing capital and fuel costs, nuclear power would have priced itself out of the market, despite governmental help, were it not for a drastic rise in the cost of fossil fuels starting suddenly in 1973. This completely revises the economic comparison between wind power, which does not require fuel, and other energy sources that do. It means that large-scale wind power is an idea whose time has finally come. Experience has shown it to be technically feasible, in need only of detailed engineering design and industrial initiative to be deployed on a large scale. It promises to be economically feasible as well. This has not yet been recognized by an appropriate revision of priorities. Energy planning statements widely recognize that its time will come, but that time is still officially declared to be in the next century—for no valid technical or economic reason. Even if the growth of nuclear power is recognized as being unwise, it may be politically difficult to stop the flow of public funds into nuclear power and to initiate large-scale wind power soon.

Recent serious discussions of the energy problem have increasingly recognized that there are limits to growth. Biological systems are subject to diseases, such as cancer, in which the control mechanism that limits cell division is lacking and uninhibited internal growth kills the organism. The human splurge of energy consumption has led to a growth economy and growth psychology in which success is measured by growth until growth itself threatens to consume the organism, in this case the global ecosystem. Various "scenarios" or "paths" for the future national energy policy are being discussed in relation to this global problem. They range from schedules which continue the historic trend of profligate and wasteful energy use to those which taper off to zero energy growth or even reduction of energy consumption, but with emphasis on achieving more good living per unit of energy consumed.

It has been suggested (Lovins, 1976) that, in the latter or "soft path," there should be less dependence on large electric distribution networks from central power stations and more on local power sources. Fuel consumption would be gradually reduced as solar-related sources would increasingly generate electricity and be used directly for tasks not efficiently served by conversion to electric power. Proponents of this view sometimes prematurely denigrate large-scale alternative power sources that depend on electric power transmission. It is reasonable to point to localization of power sources as an important contribution and an ideal replacement, but our basic manner of living and our population distribution pattern of predominantly city and suburban dwelling is completely dependent on the availability of large-scale electric power that cannot be replaced quickly with localized sources. Much can be accomplished by eliminating waste in our living pattern, and demand is still greater than need, but the need for high-line electric power remains.

If it is not supplied by large-scale wind power and other solar-related or geophysical sources, it will be supplied by less desirable sources that consume fuels.

It is an article of faith among nuclear power proponents, usually recited in the context of the historic-growth scenario, that there is no alternative to nuclear power in the next few decades. The main thesis of this book is that there *is* an attractive and economical alternative to most or all of it, starting with large-scale wind power in remote windy regions to be supplemented later by other alternative sources. This is not obvious and indeed may be considered by some to be preposterous, so more than the bare statement of the fact is needed to be convincing. Supporting details are supplied: descriptions of evolving wind-power technology, which is itself a fascinating subject, along with its prospective economics and a comparison of its advantages and disadvantages with those of nuclear power; a discussion of the present state and prospects of alternative small- and large-scale power sources; and observations and suggestions concerning current development programs and evolving political trends and possibilities. Fossil-fuel sources, on which we now mainly depend for our power, are taken for granted and not described here; nuclear power technology, the subject of an earlier book by the author, is touched on only lightly.

Given our nation's need to reduce oil imports, and given a finite globe needing to save some fuels for important future uses, it is irresponsible not to be giving top priority to the rapid introduction of alternative energy sources, along with conservation. The energy challenge should be faced with the realization that the main obstacle to generating a large share of our needed electric power from the wind soon and a bit later from other solar sources is not technical or economic, but lies rather in the way congressional appropriations and business decisions are made. They seem to be made primarily to perpetuate existing institutions and practices, not to initiate new ones.

The author is deeply grateful to his colleague Professor William E. Heronemus for initiation into an appreciation of wind power and continued tutelage therein. He is likewise thankful to his physicist colleagues, Professors Leroy Cook and William Mullin, for constructive criticism of the manuscript. Friendly discussions with members of ERDA's Solar Energy Division and Wind Power Conversion Branch have also been very helpful.

Contents

1

Past Experience
with Wind Power

For most of the million years or so of mankind's existence, his only source of power has been human muscle. Then, through most of recorded history (and a few thousand years before) man used three other sources of energy: fire for heat, domesticated animals for land transportation and other tasks, and wind to propel his boats. The use of wind power and water power on land to run mills and pumps was added very early in recorded history. The rise of European civilization from the Middle Ages through the first century of the Industrial Revolution was dependent on increased use of these power sources.

Only in the last two hundred years, since the invention of the steam engine, has the burning of fuels been used to supply power as well as heat. Most of the early steam engines were fired by wood, continually renewed by sunlight, but that has been superseded by coal, oil, and natural gas as the earth's primary power sources. These fossil fuels, so plentiful for a time and easy to extract, were formed from the energy of sunlight over a period of several hundred million years very long ago. In the last quarter century another nonrenewable fuel, uranium, has begun to enter the picture.

The size, speed, and refinement of sailing ships, propelled by the wind increased through recorded history and played a leading role in the development of history—in the growth of ancient empires, in the discovery and conquest of continents, in the migration of peoples and the spread of western civilization. Toward the end of the great age of commercial sail, and as steam gradually took over, American ships, such as the *Yankee Clipper* were the best in the world. Remnants of commercial sail still survive in parts of the world where labor is cheap and there is the possibility that it will be revived in modern dress as fuels become scarce (fig. 1). Smaller sailboats still remind us of the power of the wind.

Fig. 1.   The Dynaship. (*Courtesy DynaShip Corporation, Palo Alto, Calif.*)

The Power of the Wind

As wind power was harnessed in the past, people probably seldom asked whether the increasing number of windmills would slow down the general circulation of the wind or whether there was power to spare in the wind. Beyond the immediate wake of a windmill or the lee of a ship, man's efforts had no apparent effect on the strength of the wind. As we consider applying modern technology to feed our greatly expanded present and future use of power, one might ask the question anew. Again, the answer is that even if we should expand the past use of wind power a hundredfold or more, there is still plenty of power in the wind to spare.

Wind power is of course a form of solar power, or "solar-related power'' as it is sometimes called. The earth's atmosphere is a marvelous solar-driven heat engine. Much like the steam turbine in a modern power plant, it converts heat energy into mechanical energy by transferring heat from a warm or hot source to a cooler sink where the heat is dissipated. The global circulation of the atmosphere occurs in five rather distinct zones: the tropical zone near the equator, the two temperate zones, and the north and south polar zones. The circulation in the temperate zones conducts an enormous amount of heat from the solar-heated tropics to the radiation-cooled polar regions in an interesting and complicated way. It will be discussed further in appendix 1. Even though this tremendous heat engine is inefficient, the amount of mechanical power produced in and dissipated by the winds is a thousand times mankind's total use of power. Large numbers of wind turbines spread out over great areas would only have the effect of increasing the surface drag on the wind, similar to that of great forests. The deforestation and subsequent partial reforestation of the northern tier of states in this country made such a change in surface drag, yet it had no known effect on the general circulation of the wind. This is perhaps as direct an observation as we have that there is plenty of energy to spare to supply all the wind power we can harness in the future.

Early Wind Power Development

The use of crude windmills to pump water and to power simple machines dates back a long time in human history. There is a picture of a windmill on a Chinese vase of about the third milleneum B.C., reference to a windmill in Babylonia in the seventeenth century B.C., and one in Egypt in the third century B.C. Windmills seem to have been in common use in Persia in the seventh century A.D. and had spread to Western Europe and England by the twelfth century and to China by the thirteenth. The early windmills in Persia and China used a horizontal wind wheel carrying sails and rotating on a vertical shaft, in some cases with half of the wheel sheltered from the wind. It could turn a millstone to grind grain without gears. Modifications of this type also appeared in Europe and America much later.

The more prevalent type, with a large vertical wind wheel on a horizontal shaft, first appeared in Western Europe in the twelfth century and found increasing usage right into the nineteenth century when the invention of the steam engine provided a more convenient source of power. Perhaps its most famous and picturesque use was in sixteenth- to nineteenth century Holland, where many windmills pumped water to maintain the integrity of land reclaimed from the sea. In many countries they served a number of other purposes, mainly grinding grain but also powering machinery in sawmills, paper mills, and the like.

Through the centuries there were several improvements in design, an important one being the fantail invented in England in the eighteenth century. This is a small auxiliary wind wheel mounted at right angles to the main wheel and so geared as to keep the main wheel pointing into the wind automatically (figs. 2, 3, and 4). Until quite late in their development, windmills were made of wood, some of them with a pivoting cap surmounting a masonry tower. Throughout most of the nineteenth and twentieth centuries, smaller, many-bladed windmills mounted on slender steel or wooden towers have pumped water and have been a familiar sight on American farms.

Fig. 2. The Shipley Mill at Sussex, England, built as recently as 1879 for £2,500 and once owned by Hillaire Belloc. (Beedell, 1975.)

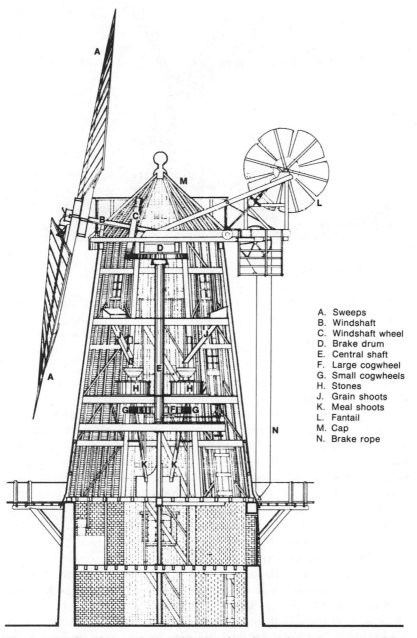

A. Sweeps
B. Windshaft
C. Windshaft wheel
D. Brake drum
E. Central shaft
F. Large cogwheel
G. Small cogwheels
H. Stones
J. Grain shoots
K. Meal shoots
L. Fantail
M. Cap
N. Brake rope

Fig. 3. Details of the Shipley Mill. The fantail *L* drives the yaw as might a much larger vane. (*Courtesy Mr. Bowles, Brighton Polytechnic, Sussex, England.*)

Fig. 4. An old print of three English windmills with fantails.

When electric lights and appliances came into common usage in the early twentieth century, many farms were too far from cities to have this convenience and farm wind-electric generators became popular. These were designed to rotate fast to drive an electric generator, and usually with only two blades so the blades would not follow each other in too quick succession, thus making best use of the wind. They could generate about 5 kilowatts, could be mounted on more slender poles or towers than the pump windmills because of the fewer blades, and could be used not only to power appliances directly but also to charge batteries for use in windless periods. Such windmills producing electric power were a common sight on American farms until the mid-thirties when, partly as a measure to relieve unemployment in the depression, rural electrification was expanded to reach most farms. This made it more convenient to use cheap electricity produced by burning cheap coal than to maintain a wind generator and storage batteries.

As for its name, the fine old word "windmill" refers to a machine of many uses. Though in origin "windmill" refers to a real mill to grind grain or to saw wood, it has long been applied to windmills that pump water and now can mean one that generates electric power. However, the modern machines that generate electricity have such a different form and function that a more specific name is desired. Several are in use. In American government reports they are known as WECS, meaning wind-electric conversion systems, and in some European writings WEGS, or wind-electric generators

—the shorter term wind generator is unacceptable because they do not generate wind. Wind turbine is a good term when the emphasis is on the part being driven by the wind, for the word turbine implies rotary motion driven by a passing fluid. There are steam turbines, compressed air turbines, water turbines, and wind turbines. But that does not imply the electric function. The single-word expression for an electric generator is dynamo. When the emphasis is on generating electric power, we use the shorter term wind dynamo in place of wind-electric generator.

Throughout the twentieth century a few people have nurtured and explored the idea that much larger wind dynamos than the ones on farms could feed power into the commercial electric power grids and thus reduce dependence on generating plants that consume fuel, with all their unpleasant implications. Denmark began even before the turn of the century and was perhaps the first country to use appreciable numbers of wind dynamos, able to produce as much as 100 kilowatts each, to generate electricity. They were a boon to that country when fuel supplies were cut off during the Nazi occupation of World War II, as were windmills to the Germans during the latter part of the war when fuel was very scarce.

In the Soviet Union, an experimental machine rated at 100 kilowatts in a twenty-four-mile-per-hour wind was built in the Crimea, in 1931. Smaller machines are used in Soviet agricultural communities.

The Grandpa's Knob Wind Dynamo

The largest and most impressive demonstration of all, and one that preceded all but those earliest European efforts, was the wind dynamo built and operated in the early forties on a small mountain called Grandpa's Knob near Rutland, Vermont (Putnam, 1945) as shown in figure 5. As a result of enthusiastic promotion by Palmer C. Putnam, it was built by the S. Morgan Smith Company, of York, Pennsylvania, a turbine firm interested in developing a new product. It fed 1,250 kilowatts, or 1.25 megawatts, into the commercial power grid of the Central Vermont Public Service Company. From the company's first interest and the decision to proceed in 1939, it took only two years to complete the engineering design, build the machine, and start feeding power into the commercial electric grid. Some parts were procured early, in the race with wartime scarcities, and when the design was completed it was found that a part supporting the blades was not strong enough. It could not be replaced but it was decided to proceed, despite the risk of breakage. In 1945, when the planned testing period was almost finished, the machine did break, throwing a blade, and the project was abandoned.

As the first big, experimental machine, it was not expected to be economic and indeed cost just over $1 million including development. The experience gained was used to design a simpler model of similar size (fig. 6) and it was estimated that building a few of them would entail a capital investment of $191 per installed kilowatt, whereas $125, or 35 percent less, would be required to be economic for the Central Vermont Power Company. The intent was to use wind power as a supplement to the main base of hydroelectric power, saving some of the water power for when the wind was light, but it turned out to be 35 percent cheaper to use power produced by the cheap coal of that era instead.

Building on this experience, the Federal Power Commission, sparked by one of its engineers, P. H. Thomas, made a serious study of the potential of wind power. An experimental 6.5 megawatt wind-electric machine, 475 feet high with two rotors (fig. 7), was designed to be used in conjunction with a hydroelectric plant (Thomas, 1946). This was enthusiastically supported in some government departments and a bill was written in Congress proposing to construct such a machine. However, the bill died in committee during the outbreak of the Korean war in 1951 when Congress was distracted by that event. This was also just at the time when euphoric claims were being made for the future of nuclear power, then under initial development. This is as close as we ever came back then to having a government-sponsored entry into the field of large-scale wind power.

Postwar Experience
Several wind dynamos of a variety of designs in the 100-kilowatt range were built and tested in England in the 1950s. After extensive wind surveys of their west coast, the English seriously considered depending largely on windpower until they decided instead to go nuclear in a big way. Wind dynamos producing up to about 100 kilowatts were also built in several other countries during that period.

In Denmark, where there had been a sixty-year history of wind dynamos feeding mainly into direct-current power lines, an interesting 200-kilowatt generator was built in 1956 at Gedser on the southern tip of the Baltic coast (Johnson, 1977). It operated dependably for ten years. The special feature of this, following the experience with earlier Danish machines, was that its three blades on the windward side of the tower were braced with guy wires, as shown

Fig. 5. The Smith-Putnam 1.25-megawatt wind turbine at Grandpa's Knob, Vermont, 1941–45.

Fig. 6. Scale model of the proposed production version of the Smith-Putnam wind dynamo.

(Figures 5 and 6 from *Power from the Wind* by Palmer Coslett Putnam. Copyright 1975 by Allis-Chalmers Corporation. Reprinted by permission of Van Nostrand Reinhold Company, a division of Litton Educational Publishing.)

Fig. 7. Federal Power Commission twin-turbine design of Percy Thomas proposed in 1951. Its rating is 6.5 megawatts and its height is 475 feet. The house is to provide a scale of reference, not to suggest that the huge machine might be sited so close to a dwelling.

in figure 8. While this arrangement seems less elegant than the slender cantilevered blades of other machines, it appears to be an effective and perhaps economical way to achieve enough strength to avoid vibration troubles in these large structures. Analogously, many small sport sailboats have unstayed masts but the masts of the tall sailing ships are conventionally braced with stays. The modern tall ship that may soon bring commercial sail back to the high seas (Wallace, 1976) will have free-standing automated masts and sails in order to save deck-hand labor (see fig. 1), but wind turbines have no such reason to avoid the lighter construction with stays.

In France, in about 1960, the development program of the nationalized utility Electricité de France went directly to much larger three-bladed wind dynamos. The program developed an 800-kilowatt (fig. 9) and a 1-megawatt machine, as well as a few smaller machines. The experiences with these were promising and plans were instituted to build more large machines with two or four rotors, but the drop of oil prices in 1963 led to termination of the program and neglect of a needed repair on the 1-megawatt machine.

In West Germany, a 100-kilowatt wind dynamo was built near

Stuttgart with special attention to the aerodynamic excellence of its slender fiberglass composite blades (fig. 10). It fed into the electric grid from 1959 to 1968. Based on experience with that machine, continued interest led to an interesting improved design concept that is discussed further in chapters 3 and 4.

All of these machines, which were successful technically if not economically, were of the conventional horizontal-axis type, with the electric generator geared to the rotor axis aloft. When considering which future options should be pursued most vigorously, it is noteworthy that some trials in France of small wind dynamos of radically different design (including a vertical-axis machine and machines with rimmed rotors) were disappointing. There was likewise a 100-kilowatt British machine of elegant concept and streamlined appearance (fig. 11) that was disappointing in its relatively high cost and low efficiency. The blades of the main rotor were hollow and as they rotated air was thrown outward through them by centrifugal force, escaping through holes in their tips and creating a partial vacuum which drew air into the closed tower. This transmitted the power to a more compact and rapidly rotating wind dynamo in the base of the tower, operating on the excess of exterior over interior pressure there. This had the advantage of reducing the

Fig. 8. Blade structure of the 200-kilowatt wind dynamo at Gedser on the southern tip of Denmark, on the Baltic sea coast, known as the Gedser Mill.

Fig. 9. Eight-hundred-kilowatt (BEST-Romani) wind dynamo at Nogent-le-Roi (Eure et Loire in northern France).

Fig. 10. Hütter's 100-kilowatt wind dynamo that operated near Stuttgart from 1959 to 1968. As compared with the Smith-Putnam machine, it had 8 percent of the rated power in a wind about half as fast, its blade diameter of 34 meters or 112 feet gave it a swept area 40 percent as great and it weighed about 6 percent as much.

weight aloft but was inefficient because it used one wind turbine to create the wind to drive another wind turbine—a windmill within a windmill, so to speak.

In Britain there was a promising early design for a large wind dynamo which, like the contemporary Federal Power Commission design in the United States, was never built. As shown in figure 12, its support is a tripod with a broad base that reduces strength requirements, in this respect following the precedent of the Soviet 100-kilowatt wind dynamo built of crude materials in the Crimea in 1931. Its two downwind legs run around a circular track on railway trucks for yawing and may be spread apart to lower the structure near ground level for construction and major servicing. Yawing is driven by a small fantail though the tripod is almost self-steering without it, swiveling about the base of the third leg (Golding, 1956).

Most of these fairly large experimental wind dynamos have had rotors with two blades, although there have been larger French machines and others with three and at least one with four blades. Practically all have successfully synchronized their alternating-current output with commercial grids, though the early Danish machines produced direct current. Most have made use of a conventional gearbox to step up the rotational speed of the turbine, on the order of 100 revolutions per minute, to the speed of an electric

Fig. 11. British 100-kilowatt Enfield-Andreau wind turbine at St. Albans. (Golding, 1955, 1976.)

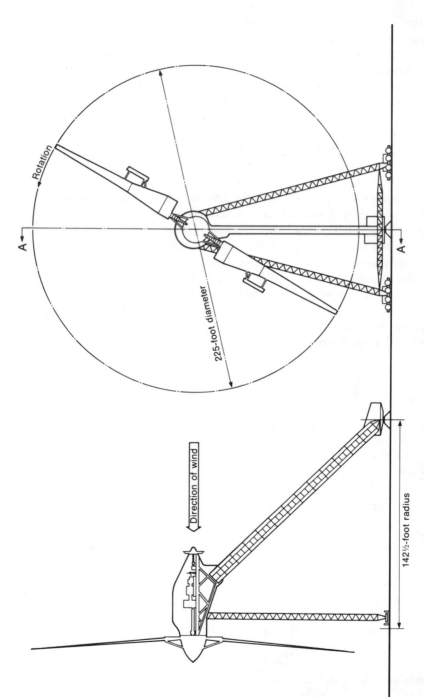

Fig. 12. British 1951 rotatable tripod design, rated at 3.65 megawatts in 35-mile-per-hour wind. Blade diameter is 225 feet. It was designed by Folland Aircraft Ltd. for the Ministry of Fuel and Power. Two legs are supported on railway trucks on a circular track pivoting about the base of the windward leg.

generator of reasonable size, usually on the order of 3,000 rpm. The gearbox is, indeed, an appreciable fraction of the cost and of the weight aloft.

Some of the wind turbines, including the Smith-Putnam and some of the British ones, had blades of constant cross section and were not twisted for reasons of economical construction. Considerable improvement in efficiency can be achieved by twisting the blade so that it is almost flat to the wind near the tip, where the speed in rotation is fast, and almost edgewise to the wind near the hub, where the speed in rotation is slow. Some of the later French and German machines had fiber-and-plastic blades of this type.

Most modern wind dynamos have blades with variable pitch. This requires an intricate hub structure in which the roots of the blades rotate, similar to the pitch-changing hub in an aircraft propeller. Not only can the turbine speed thus be automatically controlled, helping to match line frequency, but also the blades can be feathered to the wind in a storm, with the rotation stopped. Most of the British wind turbines and some others used a fixed pitch to avoid this expense. In some of these a spoiler was used, a small and narrow blade ahead of the leading edge of the main blade, to reduce the turbine's efficiency at higher speeds and help keep the power nearly constant. The tripod design of figure 12 employs variable-pitch ailerons on the trailing edge.

The area swept out by the blades is not necessarily a disc in a flat plane, but instead may be a shallow cone. That is to say, each blade need not be exactly at right angles to the shaft, as centrifugal force would tend to make it, but may lean slightly downwind as if yielding somewhat to the direct force of the wind. The adjustment to permit this variation of angle is called coning.

As is clear in the picture (fig. 5), the Smith-Putnam or Grandpa's Knob machine has in its hub structure what amounts to a huge hinge with a bracing spar supporting each blade and permitting coning. Each of the two blades can respond to the wind power of the instant and they can take on their momentary coning angles independently. While this relieves strain, it also creates coning vibrations. Such vibrations are moderated by a suitable damping mechanism. This feature is especially important for a machine like the Smith-Putnam one that has a relatively low tower compared to its blade span. In that machine the highest point reached by a blade tip, 212 feet above ground, is 5.7 times as high as the lowest point, 38 feet. Since the wind is slower near the ground, this means that the blade's wind speed differs greatly as it rotates and thus the blade has a strong tendency to vibrate with a change in coning angle. Some of the European machines with higher towers relative to their blade spans are subject to less violent variation of the wind force and do without adjustable coning. In most machines the blades are downwind from the tower and the blades feel a similar variability in the force of the wind because of the wind shadow of the tower. In some of them the tower is slender in order to minimize this effect.

All these technical problems of the stability of large wind dynamos and of synchronizing their output with the alternating current of power lines have been solved. The experience with all these machines leads to the conclusion that wind power is technically ready for broader application. One wonders why it was not pursued further.

Although the early euphoric and unwarranted expectations of nuclear energy probably had something to do with it, probably the primary reason for the lack of interest in developing wind power through the fifties and sixties has been the cheapness and abundance of fossil fuels. This reason is no longer operative and the government, mainly through the Atomic Energy Commission (AEC), and its successor organization the Energy Research and Development Authority (ERDA), and subsequently the Department of Energy (DOE), has belatedly in the mid-seventies shown increasing interest in supporting a relatively slow but accelerating development of wind power.

Another main reason for neglect of wind power seems to have been that there is almost no one in or near government decision-making circles with a strong professional interest, whereas those with a strong vested interest in rapid development of nuclear power wield very great influence and seem to resent the possibility that competing alternatives might absorb an appreciable part of available funds. This is one aspect of the conservatism of decision making. It is easier to carry on with already established programs than to start something new. Well-funded projects have the means to influence decisions and obtain more funds, or, as the saying goes, those who have, get. Forces acting on government decisions from outside also have the same built-in conservatism that tends to preserve and expand existing activities at the expense of innovation. Present activities support actual employment and thus have a political constituency. A future wind-power industry to replace some of the present power industries might hold the prospect of more jobs but they are hypothetical future jobs. No individual worker holds one or can be certain he would get one. Thus wind power has no political constituency in labor. Although capitalism and private enterprise are renowned for their achievements of innovation in the past, in the energy field capital is now attracted to established enterprises partly because of special incentives such as tax allowances that have helped establish those enterprises. The balance of competition in the energy field has been warped by government promotion and subsidization of established energy sources.

As a result, industry awaits government initiative to organize and finance the early stages of large-scale wind power rather than to forge ahead independently as did the S. Morgan Smith Company in 1940. The government demonstration program, after expanding quickly from a small beginning, is funded at about 2 percent of the level of nuclear power. It is promoting a fairly rapid spread of small-scale wind power and a wide diversity of investigations, but fostering only a slow growth of large-scale wind power, which is not expected to make a substantial contribution to the national electric power supply until the next century. The design and demonstration program is sequential, demonstrating one type of large windmill after another, rather than several types at once so as to expedite selection between them for early mass production. Such shortcomings of the present program, as well as its accomplishments, and the need for drastic expansion in the near future are discussed further in chapter 10 and appendix 4. There an encouraging recent example is cited of one small company, again, forging ahead independently in megawatt-scale wind power.

2

The Mechanics of Wind
Energy Conversion

Mechanical contrivances do not create or destroy energy but merely convert it from one form to another. One speaks of the conservation of energy. A wind turbine converts the kinetic energy of the wind's motion to mechanical energy transmitted by the shaft. A generator further converts it to electric energy, thereby "generating" electricity. A windmill is a machine for wind energy conversion.

When the wind encounters a surface, such as the blade of a windmill or the sail of a sailboat or iceboat, it exerts on it a force that can do useful work, and the process is much the same in all cases. An iceboat is more efficient than a sailboat because its runners encounter very little friction to oppose forward motion and yet normally do not skid and are more effective than the keel of a sailboat in preventing sideways motion (fig. 13). The iceboat headed at right angles to the wind is thus confined to move along a straight path in that direction in much the same way that the blade of a windmill is confined to move in a circular path at right angles to the wind. The effect of the wind is the same on both. It is simpler to think about the effect on an iceboat sail because it is easier to imagine a person sitting on the iceboat and feeling that the wind seems to come at him, not from a right angle but from a more forward direction. Most people have experienced this not on an iceboat but on a bicycle or in an open automobile: on a still day the wind, due to motion, seems to come from straight ahead, but with a cross wind the wind is felt slanting from the side. With forward speed to the west equal to wind speed from the south, for example, wind is felt coming from the southwest at 45° from forward. As forward speed increases, it is felt coming from more nearly forward. The way the wind due to motion and the actual wind com-

18

Fig. 13. An iceboat can move much faster than the wind, as do the blades of a modern wind turbine. (*Courtesy Lockley Newport Boats, Inc.*)

bine to make the effective wind as felt on the iceboat is indicated by the triangle of three arrows shown in figure 14.

An oversimplified way to explain how an iceboat (or a sailboat) works is to think of the sail as flat and with no friction from the air rubbing past it. The wind blows slantwise on one side of the sail and makes the pressure on that side greater than on the other side. The difference of pressure makes a force at right angles to the flat sail and the sail is set so that the force is partly forward, partly sideways. The forward force propels the boat. Such a flat sail would work this way but not very efficiently because at the forward edge of the sail the wind has to turn a sharp corner as it moves around behind the sail. This makes the wind motion behind the sail turbulent. In this confused motion, the wind moves relatively slowly across the back of the sail, does not get out of the way fast enough, and lets a pressure build up behind the sail almost as great as that on the side that the wind hits directly so that there is only a small pressure difference and a weak push on the sail. (We call the side of the sail away from the wind the back side although it is the side toward the front of the iceboat.)

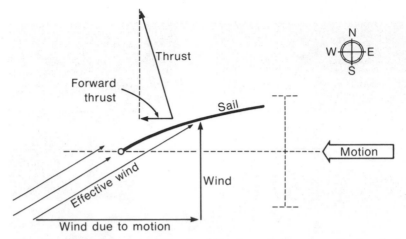

Fig. 14. The force exerted by the effective wind on the sail of an iceboat moving faster than the wind.

To work efficiently the wind must flow smoothly around the back side of the sail and this is accomplished by having the sail bent and by setting it so that at its forward edge it is parallel to the effective wind and yet the wind impinges on one side of the sail because the sail is curved. The wind encounters no sharp corner and can flow smoothly around the back side. To do so, the wind travels on a curved path, which requires that there must be a force pushing sideways on this stream of air to bend it away from a straight path. This force is provided by pressure being less on the inside of the curve than on the outside of the curve, meaning that the pressure is reduced on the back side of the sail. In the same way the stream of wind on the front side of the sail is being pushed into its curved path by an increase of pressure on the front side of the sail. This increased pressure on the front as compared with the decreased pressure on the back is very effective in exerting a thrust on the sail almost at right angles to the general direction of the curved sail, as indicated in figure 14. This thrust is slanted somewhat forward and thus creates a forward thrust that propels the iceboat. An iceboat has very little friction so the forward thrust need not be strong and can be almost at right angles to the direction of motion. The sail can thus be nearly parallel to the direction of motion so the effective wind can come almost from straight ahead and still be parallel to the sail at its forward edge if the sail is curved only slightly. The wind due to motion can then be quite a bit greater than the actual wind, allowing the iceboat to sail considerably faster than the actual wind.

Fig. 15. Some of the more than ten thousand windmills that pump irrigation water in Crete. They use sails much like sailboat sails. (Copyright 1978 National Geographic Society.)

Some windmills have actual sails for blades, like the windmills in Crete (fig. 15). A type is now being developed in the United States called the Sailwing. The above discussion applies just as well to sailwings as to an iceboat. The tips of the blades or sails of a windmill can similarly move much faster than the wind.

Early attempts to build machines that would fly employed thin wings, at first flat, then slightly curved to resemble the curvature of a sail. Real progress in the technology of aviation began with the invention of the airfoil, the curved shape of a thick wing, which greatly increased the lift obtainable from a given area of wing. Airplanes and modern windmills have much in common: they employ the same processes, but in a different order. In the case of an airplane, the power of the fuel-burning engine creates the wind, which in turn creates the desired lift. Thus the sequence is from power to wind to lift, whereas for a windmill it is from wind to lift to power. The wind provided by Nature creates the lift or thrust, and the lift creates the desired power. In the airplane, the power creates a wind backward on the wings by propelling the airplane forward. An airfoil, such as shown in figure 16, differs from a sail, being thick near the leading edge and tapered down to a thin trailing edge. This means that the curvatures of the bottom (or back) surface and the top (or front) surface are not the same and each can be designed to perform its function best—the bottom surface to allow an increased pressure on it and the top surface to allow a decreased pressure there. It permits the top surface to have a rather sharp curvature near the leading edge without creating a cavity in the lower surface that would have a sharp forward edge and would induce turbulence. The air flowing around the sharp curve on the fore part of the upper surface experiences a centrifugal force, so to speak, and is very effective in decreasing the pressure on the wing there and further aft. The gradually decreasing curvature then guides the air to flow smoothly back to the trailing edge. The air as it curves downward along the top edge of the wing presses harder on the air above it, outside the curve, than it does on the air below it and on the wing itself. Because of the centrifugal force as it goes around the curve, it might even be said to pull the wing upward, meaning that the air pushes down on the top of the wing less than the higher pressure air below the wing pushes it up.

A consequence is that the flow is faster over the top surface than over the bottom surface. The air is speeded up, pushed by a pressure difference, coming in from the high-pressure region ahead of the wing to the low-pressure region along the top surface. This

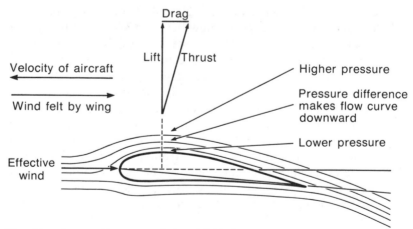

Fig. 16. Stream flow around the airfoil of an airplane wing.

is in keeping with a principle that is attributed to Bernoulli: where the pressure is least, the speed is greatest. Behind the trailing edge, where the fast flow from the top surface joins the slower flow from the bottom, the combination of the two streams curves downward in eddies that break off and are left behind. The eddies carry momentum downward and the stronger the eddies, the greater the lift on the wing.

Horizontal Axis Windmills

The blade of the windmill may have either the thin cross section of a sail or the more efficient thick cross section of an airfoil as suggested in figure 17, which is similar to figure 16. The motion causing the "wind due to motion" here is the rotation of the blades. At the tip of the blades of a modern wind turbine that velocity is about six times the wind velocity. This means that the blades are set rather flat at a small angle with the plane of the rotation and almost at right angles to the direction of the wind so that the effective wind will properly approach from ahead of the leading edge. At other parts of the blade, between the tip and the axle, the velocity produced by the rotation is not so great as six times the wind velocity and the ideal set of the airfoil is at a greater angle to the plane of rotation. Ideally the blade should be twisted, but because of construction difficulties this is not always achieved.

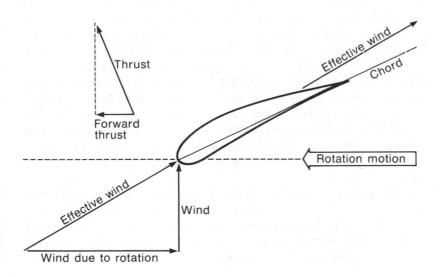

Fig. 17. Windmill blade as an airfoil.

Fast and Slow Windmills
Two types of windmills have been familiar on American farms;
rapidly rotating wind dynamos generating electric power and more
slowly rotating windmills pumping water. An electric generator is
more efficient if designed for rapid rotation, and even with rapidly
rotating windmill blades, the rotation is speeded up by gears or pul-
leys driving the generator. On the other hand, a slow reciprocating motion from a crank is more compatible with the needs of a
conventional water pump. The fast electricity-generating wind
dynamo has only two or three blades while the pumping windmill
has a dozen or more (fig. 18).

When the blade of a rapidly rotating windmill passes a certain
point in the arc, say the topmost point, it disturbs the air for a
short distance both upstream and downstream from it. If at its tip
the blade is moving at about six times wind velocity, this means that
the air stream will have moved only one-sixth of the distance measured around the arc between blades before the next blade cuts
across the same airstream. This is about as soon as it can come into
almost undisturbed air so as to be able to extract as much power
from the next section of the airstream. In this way the rapidly rotating blades, even though their solid surface occupies only a small
part of the circle they traverse, extract power effectively from the
entire circle. They slow down considerably the air mass passing
through the circle, and the reduction of the kinetic energy of the
air mass appears as the power produced by the windmill. Of course
it cannot quite stop the air mass, and the theoretical limit of the
power that can be extracted amounts to about 60 percent of the
incoming kinetic energy, as is discussed at the end of appendix 2.

A slowly rotating, water-pumping windmill as shown in figure
18, on the other hand, has its blades quite close together in order
for one blade to cut in rather soon behind another and make effective use of the power of the wind passing through its circle. This
means that the material surface of the blades covers a large part of
the area of the circle. If a very large windmill, as large as the one at
Grandpa's Knob, were built this way it would require an exorbitant
weight of material.

Some of the older windmills of previous centuries have such
wide blades that, with only four or six blades, a considerable fraction of the area of the circle is covered and they harness the energy
of the wind passing through the circle more than half as efficiently
as a modern high-speed wind turbine does.

Fig. 18. A farm windmill with storage tank for watering cattle in northern Arizona.

Most of the successful wind turbines have some arrangement to adjust the effectiveness of the blades. In some the effective area of the blade or sail is varied, and in others the pitch of the blade. This adjustment serves a dual purpose. It is used to control the speed or power output; for example, to keep the speed constant as the wind varies. It also may help the structure to survive a storm by presenting as small an area as possible to the wind.

On the familiar water-pumping farm windmill this is done by changing the angle between the blade circle and the vane that points downwind. The farmer pulls a lever to set the blade circle edgewise to the wind and stop the rotation in a very high wind. One type of windmill popular in the Mediterranean, shown in figure 15, has six or eight poles that carry sails the way the mast of a sailboat does, and the sails may be reefed by rolling them on the poles or some of the sails may be removed. The big old European windmills have various reefing arrangements. One type used particularly in England (fig. 2) has the blade covered with transverse slats. The rotational speed is kept fairly constant by automatically letting the slats tilt so as to spill some of the wind through the blade. In tilting they pull either againt springs or against a counterweight to which they are tethered through a hollow main shaft. Many old windmills had lattice frames for blades on which sails were spread, as in figure 4. The sails were reefed to present the desired area to the wind by rolling them along one long edge. The miller did this by twisting one end of the sail as it was stopped pointing downward, hardly an automatic device. Some fairly modern wind dynamos, like the one in figure 8, have blades of fixed pitch but with a rotatable flap at the tip end of each blade. This saves the expense of an intricate hub but does not provide for feathering in a storm. Machines of this type in Denmark have nevertheless survived for decades.

Most modern high-speed wind turbines with two or three blades have a mechanism in the hub that permits changing the pitch of the blades, as is done with some airplane propellers. In a storm the blades can be "feathered" to the wind.

Pivoting to Face the Wind

Normally the wind keeps changing its direction and provision must be made for a horizontal-axis windmill to turn on a vertical axis to keep it facing approximately into the wind. This turning about a vertical axis is called "yawing." This is unnecessary only in a few special locations such as mountain passes or certain islands where the wind almost always blows from the same direction. Various contrivances are used for accomplishing this. Modern windmills can be pivoted on good bearings making the turning easy, so a simple wind vane, mounted to be swept downwind, suffices. This is seen on most farm windmills, both the slow pumping type and the rapid type generating electric power. The same effect may be accomplished without the vane by mounting the blades on the downwind side of the pivot and generator and having the blades coned, or bent a little away from the wind so that in rotating they sweep out a broad cone rather than a flat circular area. Thus the blades, being on the downwind side, perform the function of the vane.

In giant electric-generating windmills a more positive and controlled orienting effect may be achieved by using electric power, monitored by a simple weather vane to do the turning—a sort of power-steering arrangement. This was done, for example, in the Grandpa's Knob machine, which also had the blades on the downwind side. Incidentally, the axis about which its blades rotated was not quite horizontal but dipped a bit downward over the brow of the hill to conform to the rising trend of the prevailing wind.

The possibility of building windmills as floating structures anchored at sea or on lakes is discussed in chapter 4. One incidental advantage of this scheme is that the yawing to face approximately into the wind would be provided automatically by riding at anchor, the way an anchored boat rides with its bow into the wind.

Gyroscopic Action
The rotating blades of a windmill act as a large gyroscope and this places a limit on how fast the rotor may be pivoted about a vertical axis. This yaw rate is limited by the strength of materials; therefore, the blades, shafts, and bearings must be designed with this requirement in mind. Everyone who has played with a gyroscope knows how ornery it is; when one tries to turn it one way it turns some other way instead. Consider a big wind turbine as a gyroscope, free to spin about a horizontal axis but also pivoted about a vertical axis. If, on a windless day when it is not spinning, one wanted to shift the axis from an east-west direction to a north-south direction, a torque about the vertical axis would be applied. When the rotor is spinning, on the other hand, as this same pivoting is started from east-west to north-south, the bearings of the vertical axis have to exert a strong torque about the north-south direction. This torque must be transmitted through the horizontal axle and its bearings to the blades and also by the roots of the blades near the hub to the main mass of the blades. For this and other reasons, all those members must be strong. It is helpful to have the blades made of a light yet sturdy material. The lighter the gyroscope spinning at a certain speed, the easier it is to pivot it. The Grandpa's Knob blades were of stainless steel, but composite fiberglass and epoxy blades are better.

A windmill must in any case be designed to be strong enough to survive a severe storm, even a tornado or hurricane. If a windmill is sturdy enough for this, it is presumably strong enough for reasonably rapid pivoting in normal operation, but it is important not to encounter both problems at once. Excessive centrifugal force must be avoided as well. This can be avoided by stopping the spinning and foregoing power production when a bad storm develops, by feathering the blades and applying brakes.

Vibration Problems

Rotating machinery is subject to vibrations and the rotating blades of a horizontal-axis wind turbine undergo special types of changing forces that induce vibrations in them and their supports. The variation of wind strength with height means that a blade experiences a stronger wind force at the top of its arc than at the bottom. This tends to make the blades vibrate relative to each other and relative to the axle. The vibrations can involve slight changes both in the coning angle and in the angle between blades. If the rotor is on the downward side of the tower, as it is in most large wind turbines, the wind shadow cast by the tower can be a more troublesome source of such vibrations because of its sudden impact. With the rotor downwind, the tower should be designed to minimize wind shadow as well as cost.

When the wind is gusty and changes direction so rapidly that the yaw mechanism cannot keep up with it, the wind approaching the rotor at an angle will exert quite a different force on the blades at the top and bottom of their paths because it makes different angles with their chords. This difference also induces vibrations. The source of vibrations in helicopter rotors is very similar; the blades traveling forward on one side and backward on the other experience rapidly varying winds. Some of the techniques for meeting this problem in helicopters and equilizing the lift on both sides may be applicable to wind turbines.

A two-bladed rotor presents a special vibration problem when its axis is being rotated to follow a change in wind direction, that is, when it is yawing. This arises from a peculiarity of gyroscopic action in the rotation of a long, slender object. It is much easier to spin a long stick about its long axis than to start it rotating about an axis perpendicular to its length. In terms used in physics, one of its moments of inertia is much smaller than the other two, almost zero. Similarly, when the two rotating blades are momentarily in their vertical position, it is much easier to drive a yawing motion than when they are in their horizontal position. Stated a bit more profoundly, there are torques of Coriolis forces affecting the yaw as they swing between these two positions. This means that the yawing motion is jerky, setting up vibrations. One engineer at Grandpa's Knob remarked that when he was in the cab beside the generator, such vibrations almost made him seasick, yet the structure was quite strong enough to withstand them.

With most types of blade construction, the cost of the blades is a considerable portion, perhaps half, of the cost of a wind dynamo. For this reason two-bladed rotors are quite commonly used and there even have been experiments with rotors consisting of a single blade and a counterweight, rotating fast enough to make effective use of the wind stream through the blade circle (and thus reducing the cost of the gear train to drive a fast electric generator). Yet avoiding serious vibrations is so important that it may well be worthwhile to go to the extra cost of three blades, particularly since there are indications that fiberglass construction may reduce their cost. Of the three megawatt-scale wind dynamos that have been built in the past, the one at Grandpa's Knob had a two-bladed rotor, the other two in France had three-bladed rotors. Present demonstrations of large wind turbines should include three-bladed rotors as well as two-bladed, for comparison of performance.

Vertical-Axis Windmills

The Savonius Rotor

As has been mentioned, probably the earliest ancient windmills rotated about a vertical axis. A grinding wheel could then be powered without gears and such windmills were also used with gears for raising irrigation water. A form still used in China and Thailand has several vertical masts on a horizontal wheel, each carrying a sail that moves around the circular path. In sailing language, it luffs and goes about on the upwind lap and jibes as it rides before

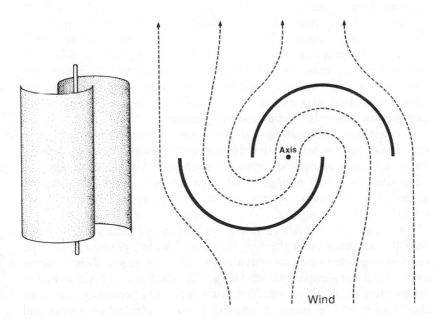

Axis

Wind

Fig. 19. The Savonius rotor and its stream flow.

the wind, thus being on the appropriate tack to drive the rotation on both the windward and leeward reaches, just as a boat might sail an almost circular course. One early type had fixed blades that were shielded through half of their rotation by a shroud that would have to be moved as the wind changed direction, but the types with sails and the more modern types of vertical-axis machines work equally well with the wind coming from any direction.

Perhaps the simplest of the modern types is the Savonius rotor which works like a cup anemometer. It consists of two half-cylinders facing opposite directions in such a way as to have almost an S-shaped cross section. However, instead of having two edges together to make an S-shape, they overlap to leave a wide space between the two inner edges, so that each of these edges is near the central axis of the opposite half-cylinder, as shown in figure 19. The main action of the wind is very simple: the force of the wind is greater on the cupped face than on the rounded face. In detail it is a bit more complicated. The wind curving around the back side of the cupped face exerts a reduced pressure much as the wind does over the top of an airfoil and this helps drive the rotation. The wide slot between the two inner edges of the half-cylinders lets the air whip around inside the forward-moving cupped face and then around the inside of the backward moving face, thus pushing both in the direction of the rotation.

Another advantage of the Savonius rotor is its low cut-in speed (the wind speed required for switching electric power into the line): it produces power effectively in winds as slow as five miles per hour, whereas most propeller-type windmills require about ten miles per hour for effective operation and large windmills still more. This means that it is useful more of the time and is thus less dependent on storage or supplementary power.

The main disadvantage of this type of machine is that it is too solid, having so much metal or other material surface compared with the amount of wind intercepted. This not only leads to excessive weight for a large installation but also leaves the machine at the mercy of severe storms, since there is no way to reduce the effective area.

In common with all vertical-axis machines, the Savonius rotor has the advantage that the weight of the electric generator may be carried at ground level without the use of bevel gears. This is, however, not a very important advantage. It is not useful for a very tall installation because a long drive shaft presents problems, and also the bracing of the topmost bearing above the rotor of a very tall vertical-axis machine is awkward, requiring very long guy wires. In a conventional horizontal-axis wind-electric machine with the generator aloft, the strength of structure required to carry the added weight of the generator is small compared with that needed to survive a severe storm, and the generator housing adds little to the area presented to the storm.

In mechanical design there are laws of scaling epitomized by the statement that a flea can jump higher than an elephant can. More seriously, as a system is made larger, it needs more than proportionally added strength to support it and to withstand the greater stresses it experiences. It is therefore more important that the larger installation have an efficient design that is economic in materials. Thus the disadvantages of a Savonius rotor do not apply to a relatively small installation that might be suitable for supplying electric power to an individual household. Indeed, it is a favorable type for this purpose, largely because of its low cut-in speed.

The Darrieus Eggbeater

As noted, a modern rapidly rotating propeller-type windmill, by use of an efficient airfoil, effectively intercepts a large area of wind with a small blade area. The Darrieus windmill is a type of vertical-axis machine that has this same advantage. An additional advantage is that it supports its blades in a way that minimizes bending stresses in normal operation.

The blades are curved and attached to hubs on the vertical shaft at both ends to form a cagelike structure suggestive of an ordinary eggbeater (fig. 20). The curved blade has the shape that a rope would take if subjected to centrifugal force in rapid rotation, something like the shape of the rope in the exercise of skipping rope. Thus the force in the blade due to the rotation is pure tension. This provides a stiffness to help withstand the wind forces it experiences. The blades can thus be made lighter than in the propeller type. What happens in a severe storm is another question, as yet unanswered.

As we have seen, the force that propels the blades of a conventional windmill comes from the chord of the airfoil, as in figure 17, being tilted away from the direction of motion, just as is the sail in figure 14, so that the thrust that is almost at right angles to the airfoil is tilted toward the forward direction and has a component in that direction, labeled "forward thrust" in figure 14. The remarkable thing about the Darrieus rotor is that it cannot have the advantage of this tilt of the airfoil and yet works without it, with the chord directly along the tangent to the circular path in the equatorial cross section shown in figure 21. If the chord were

Fig. 20. Eighty-kilowatt Darrieus wind turbine at Albuquerque, New Mexico. Its blade diameter is 55 feet. Tests indicate a power coefficient (output compared with kinetic energy brought into its area by the wind) of about 1/3, as compared with about 1/2 attained by propeller types or the Betz limit 59 percent discussed in appendix 2. One of the experimental projects of the federally funded wind energy program. (*Courtesy Sandia Laboratories.*)

tilted away from the tangent so as to tilt the thrust forward where the wind meets the airfoil on the windward side of the circle, as indicated by the broken lines in the figure, then on the other side of the circle the wind would meet the other side of the airfoil and the thrust would be tilted backward to retard the motion. (The more complicated gyromill discussed below changes the tilt on the way around.)

The reason the airfoil lie right along the circumference of the circle it traverses, but the airfoil is also symmetric about its chord for the same reason: if it were curved more on one side than on the other to make maximum thrust as in figures 14 and 17, this would be favorable going one way and unfavorable going the other way.

The reason the Darrieus rotor works at all is that the thrust is not quite at right angles to the airfoil, but slightly forward, as shown in figure 22. This comes about from a rather subtle aerodynamic effect. In figure 22 the wind is seen coming at the leading edge of the symmetric airfoil from slightly below. This means that

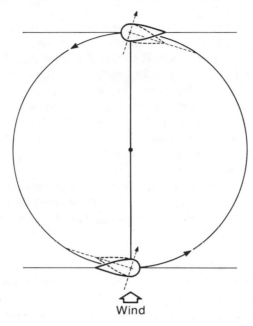

Wind

Fig. 21. Blade orientation in the Darrieus rotor. The solid lines indicate the actual position of the airfoil. The broken lines show a hypothetical orientation that would be less effective.

the wind curves around the front of the airfoil as it starts to curve around the top side. This curvature around the leading edge makes the pressure low there. The higher pressure on other parts of the wing that are not so sharply curved thus pushes the wing forward. In a manner of speaking and rather surprisingly, the wind sucks the airfoil toward the wind. There is a small higher-pressure area on the lower half of the leading edge where the stream parts. This, however, has less effect than does the low-pressure areas where the upper stream curves around the top half of the leading edge.

This subtle effect works only when the effective wind comes from a forward enough angle that the flow around the airfoil is smooth. Because the forward slant arises from the speed of rotational motion, this means that the rotor must be rotating at a required speed before the wind exerts any driving force at all on it. Thus the Darrieus eggbeater is not a self-starter. It supplies power only after some other device has brought it up to speed. In practice, the simplest way to bring it up to speed is to combine it with a considerably smaller Savonius rotor on the same shaft. The Savonius rotor is efficient at low speeds and is capable of starting the Darrieus rotor but cannot supply nearly as much power as the Darrieus rotor does at high speeds. This combination has been used in small isolated machines, but for large Darrieus wind dynamos feeding into power grids, it is more practical to use the generator that produces the output power also as a starting motor drawing on the line.

Because of the subtle way the wind drives it, the Darrieus rotor is less efficient than conventional wind turbines in extracting power from the wind passing through it (J. Carter, 1977). It seems possible that it could be made slightly more efficient by allowing enough flexibility for twisting so that the trailing edge of the airfoil could be blown slightly out of line by the force of the wind to make a favorable slant on both the near side and the far side of the circular motion. Such induced twisting might be hard to achieve in the curved blade, particularly in amounts suitable for various wind speeds.

Straight-Blade Modifications of the Darrieus Rotor
The curved shape of the blades on a Darrieus rotor provides a strong support against the strong centrifugal force with little material—employing tensile strength only—but is otherwise not essential to the mechanics of the driving force. It may be advantageous instead to use straight and approximately vertical blades of the same symmetric-airfoil cross sections of figure 22, carried on a framework to rotate about the vertical axis. They must be stiff enough or

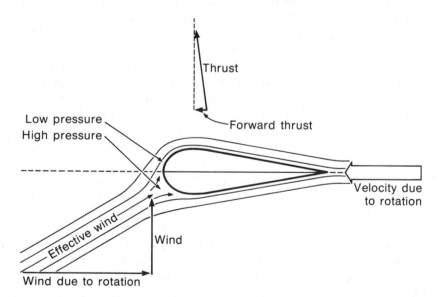

Fig. 22. Symmetric airfoil of the Darrieus rotor blade.

well enough braced, however, to withstand centrifugal force that tends to bend them. There are two interesting ways to permit adjustable orientation of such blades.

One vertical-axis machine that works on exactly the same principle as the Darrieus rotor, with the chord of the airfoil tangent to the circular path as shown by the solid-line profile in figure 21, is called the variable-geometry windmill (Musgrove, 1977). The two straight blades are mounted at the ends of a rotating horizontal crossbar and are normally vertical so as to form a shape like the letter H with the crossbar somewhat below the middle of the blades. They are hinged to the ends of the crossbar in such a way that their upper ends can tilt outward away from the vertical and they are constrained from doing this at low rotational speeds by springs. As the rotation speeds up, centrifugal force stretches the springs and the outward tilt of the blades reduces their effectiveness and tends to keep the rotational speed nearly constant. This is thus a good method of regulating the speed in small machines where simplicity is essential.

The proposed Giromill (Eldridge, 1975, p. 452; Brulle, 1976) has three vertical blades mounted on at least two three-pronged crossbars, one at the top and one at the bottom (fig. 23). As they

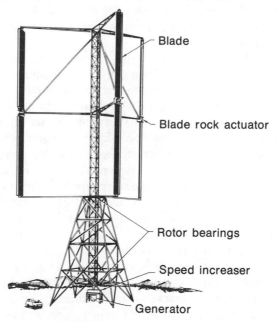

Fig. 23. Proposed 500-kilowatt Giromill. (*Courtesy McDonnell Aircraft Company.*)

swing in a circular path about the central vertical axis, each blade is rocked about its own vertical axis within the framework in such a way that at each position it has a favorable angle of attack to help drive the rotation. An aircraft company investigating the feasibility of this concept under a DOE federal contract, after extensive engineering studies of the dynamics and vibrations and some wind-tunnel testing of a two-meter-diameter model, estimates that the cost of a megawatt-scale gyromill would be competitive with similar estimates made for conventional propeller-type dynamos. Instead of having guy wires or a fixed tower inside a rotor frame, the design selected incorporates a rotating upper part of the tower mounted on large bearings within the lower part of the tower as part of the rotor itself. This means that the heavy rotor extends far beyond its bearings and its swaying vibrations are complicated by gyroscopic action. The same effect is less severe in a conventional wind turbine with a lighter rotor nearer its bearings. In the proposed gyromill design—though not in the model tested—the rocking of the blades is to be controlled by a sophisticated electronic system sensing wind direction, blade angles, and structure vibrations and commanding corrections to the blade angles. In case of vibration trouble or windstorm the blades may be set free to feather themselves in the wind. With this system the straight blades have an advantage over the twisted blades of a modern propeller-type wind turbine that cannot be completely feathered.

The gyromill seems like a good idea and is perhaps the best way to avoid the necessity of a yaw mechanism. It will be interesting to learn from full-scale test results how it compares in efficiency with a Darrieus rotor or with a conventional wind turbine. It seems unlikely, however, that its advantages will outweigh the disadvantage of so heavy and complicated a rotor when it comes to cost-effectiveness. It appears to be a more complicated and probably a more expensive way to sweep over a given area of wind than the conventional horizontal-axis propeller-type turbine and would be more subject to vibrations induced by the rapidly varying forces on its blades. Much of current vertical-axis wind-turbine research and development is concerned with analyzing and minimizing troublesome vibrations (Weatherholt, 1976).

It has not been demonstrated on a large scale but does appear to be practical in a small size. A 2-kilowatt version of this type of wind turbine is being marketed independently by a Cape Cod firm under the alternative trade name Cycloturbine (Pinson, 1977). It is designed to have a relatively short free-standing part of the vertical shaft by having two nearly horizontal support members join each vertical blade about midway between the middle and ends.

Besides not requiring a yaw mechanism, vertical-axis wind turbines have another advantage, that is, the power can be transmitted to ground level by a simple shaft without the use of bevel gears. Thus the weight of the electric generator need not be carried aloft. Carrying this extra weight aloft is not a serious disadvantage in most horizontal-axis designs because a structure strong enough to withstand high winds is more than strong enough to support the extra weight. The Hütter "leaning tower" design shown in figure 24 is an exception that achieves a less massive tower than usual and uses bevel gears aloft to drive a nearly vertical shaft and ground-level generators. Another exception is hydraulic transmission of the power to ground level, as is being developed by a Canadian institute under the name Hydrowind, and by an independent New Hampshire engineering concern—an encouraging example of recent private enterprise in wind power (Browning, 1977). In one model a reciprocating pump concentric with the yaw axis has its cylinder fixed to the top of the tower and only the piston within it turns as the wind turbine driving it yaws in the veering wind. In a much larger machine built by an engineering firm in the state of Washington, as described in appendix 4 (see fig. 57), power is

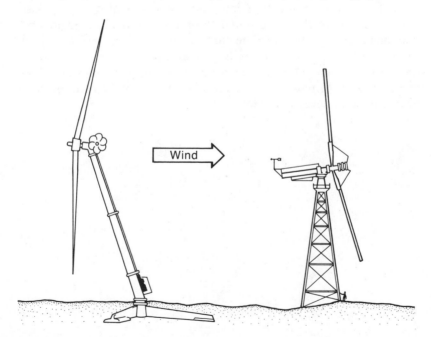

Fig. 24. Hütter's proposed 1-megawatt leaning tower of Stuttgart (*left*) compared in size with the 1.25-megawatt Smith-Putnam design (*right*). (Adapted from Hütter, 1974.)

transmitted hydraulically to a power house at ground level. This also serves the purpose of decoupling the fixed alternator speed from the rotor speed that is varied to use the wind most efficiently.

Shrouded Windmills

A low cut-in speed, or the wind speed at which a windmill begins to produce useful power, may be as important in some applications as is the rated power in fairly strong winds in determining the overall value of an installation, because low cut-in speed means minimum dependence on energy storage. One way to attain low cut-in speed is to arrange stationary shrouds to intensify the wind driving a rotor.

An obvious way to intensify the wind is to have a huge funnel in front of the rotor which will funnel the wind and concentrate the wind stream. In aerodynamic practice, however, it seems to be even more effective to use a reverse funnel behind the rotor to suck the air way and reduce the pressure behind the rotor. Shrouds have been designed using both effects but with a larger reverse funnel emphasizing the sucking effect. The shapes involve slowly curving surfaces to maintain smooth flow. To be effective, such shrouds must be quite large compared to the rotor. This means not only a serious problem of pivoting into the wind, but also a very large solidly-covered surface presented to severe storms. The engineering may be practicable because the shroud may be made very heavy and supported by wheels on a circular track directly on the ground, for the sake of orientation into the wind. Such huge masses are, however, unlikely to be practical from the point of view of both dollar economy and energy economy.

3

Small Wind-Power Installations

One nice thing about the wind is that it is available practically everywhere, in some places more steadily and stronger than in others, but, sufficient for home use for many people as well as in favored remote places. Important contributions to national energy needs may be made by perhaps millions of relatively small, kilowatt-scale windmills serving individual homes or small workshops as well as by thousands of huge, megawatt-scale wind dynamos feeding power into the commercial electric grids. The incentives for investing in the two scales of machines are quite different, but they serve the same end—providing the power on which our comfortable way of life is based. The incentive for the individual home or factory owner will be reduced dependence on expensive outside fuel and electricity supplies. Incentives for the building of large machines will be profits in producing and distributing electric power and possibly in producing other products, such as aluminum, fertilizers, and fuels.

Small windmills can be installed relatively quickly by individuals and small businesses. In the very near future their collective contribution is apt to be greater than that of large-scale wind dynamos. There are quite a lot of people motivated to do what they can to help and who will also appreciate reaping the economic benefits of harnessing wind power. Such important activity is being encouraged by ERDA-DOE programs aimed at improving the economy of small windmills.

There appears to be, however, a much larger potential contribution from large-scale wind power. Once industry perceives the economy of large-scale wind power, it seems likely that there will be several tens of thousands of megawatt-scale wind dynamos and perhaps several million households exercising the initiative to take advantage of kilowatt-scale wind power. These figures would imply roughly ten times as much power generated in large-scale as in small-scale wind power systems. There are advantages in placement

39

of energy systems with sources near points of use and with much of the energy not passing through the electric form at all (Lovins, 1976); but the electric power networks on about their present scale have become a necessary part of our way of life and residential pattern. They must be supplied with energy, preferably by megawatt-scale wind power whenever feasible. For this reason the primary emphasis in this book is placed on large-scale wind power and the extent to which it can reduce the nation's dependence on other power sources for electric power.

Home Electric Power and Home Heating

The farm wind-electric plants that were so common in the 1930s, with wind-dynamo ratings up to 5 kilowatts, were complete and independent electric systems, for there was no outside source of electric power to supplement them. They therefore included a bank of storage batteries to provide power during windless periods. These represent a substantial part of the cost of the system and require some attention for upkeep. They supply direct current rather than the more usual alternating current. This can power lights and motors, but is unsuitable for some modern electric appliances. A motor-generator can be included to convert to alternating current but at substantial added cost.

Now almost all homes are within easy reach of electric power lines, and the most common use of home wind-electric generators is as a supplement to commercial electric power for the sake of reducing power bills and ultimately saving fuel at the power plant. In this application, if alternating current is generated, there are problems of regulating the frequency and phase, and of switching from one source to the other or combining them depending on the strength of the wind. Probably the simplest solution is to generate direct current with the wind but to feed the power into the alternating-current line by means of solid-state rectifiers in a transistor circuit arranged to feed bursts of power into the line at appropriate parts of the cycle, a process known as synchronous inversion (Meyer, 1976). Thus the power is fed in at a frequency and phase determined by that of the AC line. The wave is not as smooth as that of the AC line, but has a somewhat square-cornered or "jerky" wave form that would operate most appliances poorly, if at all, were it not smoothed out as it combines with the power from the line. As long as there are not too many such wind-power installations, however, the line is capable of doing sufficient smoothing. An advantage of this scheme is that it can make use of the full power of the windmill whether or not this exceeds the power being used in the household. When the wind is light, the household draws its power partly from the wind and partly from the line, but when the wind is strong and generating an excess of power, the excess goes back into the line, helping to meet the other needs of the utility grid. The meter actually runs backward! It remains to be seen whether utility companies will endorse this practice, but it has been accepted and is in use in a few localities.

Producing wind dynamos in the 1- to 5-kilowatt range for use on farms was a thriving business in the 1930s. That business has recently come to life again in the form of relatively small companies either refurbishing old wind dynamos or starting to build new ones, both for individual home use and for small community use. The number and capabilities of such companies are rapidly growing to meet the needs of home owners interested in installing windmills. For up-to-date information about them a good starting point is the quarterly magazine *Wind Power Digest*.

An encouraging example of a wind dynamo for community use is a 200-kilowatt one built in 1977 on Cuttyhunk Island, just west of Martha's Vineyard off the Massachusetts coast (Wellikoff, 1977). It has three 40-foot blades and is rather similar in construction to the Gedser mill in Denmark (fig. 8) but with unstayed blades. The rotatable tips on fixed-pitch blades are used only as brakes. The wind dynamo is designed as a fuel saver to supplement the island's diesel generators, the load being automatically divided between the two sources by a computerlike electronic control to keep the turbine speed and generator frequency constant as the wind changes. In the recent revival of interest in wind power, it is noteworthy that a small company has put into commercial operation this wind dynamo nearly as powerful as those being constructed concurrently for community use in the federal demonstration program. These and other recent developments are discussed further in appendix 4.

Many homes are heated with electricity by use of simple resistance heating units because this form is the cheapest to install. Builders favor it because it requires none of the ducting or piping associated with a furnace and many home buyers prefer a smaller initial cost even if it means higher heating bills later. Furthermore, it has been common practice by some utility companies to encourage power consumption by giving builders a bonus for installing resistance heaters that use a lot of power.

In houses so equipped, wind power can be used as a supplementary source of heat without the complications of tying in with the power line. Even simpler than completely switching from one source to the other is connecting one source to some heating elements and the other source to the rest, with two thermostats set so that power is drawn from the power line only when the wind generator fails to keep the house warm. In this way, no storage is needed. A still more modest way to reduce power bills is to heat water using wind power. Home water heaters are commonly equipped with two heating elements with separate thermostats. The wind dynamo can be connected to one and the power line to the other, with thermostats set to call on wind first. Windpower is especially effective for space heating because it is most plentiful when the heat is needed most, on cold windy days when the cold is penetrating.

Resistor space heating with commercial electric power is an inexcusably inefficient use of our natural resources because the heat from fuel is converted into "high-grade" electric power with an efficiency of only about one-third. This may be 40 percent for the newest fossil-fuel plants but dips down to nearly 30 percent for some older plants and nuclear power plants. This means that roughly twice as much of the energy from the fuel goes into the cooling water of the power plant as goes into the electric line. Beyond that there are transmission-line losses of a few percent before the power arrives to heat the home. This inefficiency is the price that has to be paid for having the power arrive in the wonderfully convenient form of electricity, which began as heat at the power plant that boiled water to make steam that will run a turbine as long as the low-pressure steam on the other side of the turbine is condensed by the cooling water. This production of mechanical energy from heat energy is where the main inefficiency lies. The subsequent generation of electricity from the mechanical turning of the shaft is quite efficient.

Resistance heating commits the ecological sin of converting this high-grade electric energy directly back into the heat form from which it started, but only one-third as much of it. Burning the fuel directly in a home furnace, rather than at the power plant, could heat the home more efficiently, though it requires a well-adjusted furnace to be really much more efficient. Some people compensate for the intrinsic inefficiency of resistor heating by a combination of architecture and life-style that permits turning off the heat in all rooms not being used at a given time. Good insulation is of course important with any type of heating.

A resistor simply converts electric energy into heat. There is a better way to heat a house electrically, and that is to use the heat that already exists in seemingly cool substances, and pump it up to a higher temperature. This is the opposite of the familiar process that cools a refrigerator. There the heat from the inside of the refrigerator is pumped to the outside, leaving the inside cool, and the heat is dissipated in the room outside the box by a radiator that is usually under the box. If the cooling coils were put outside the house, with the radiator left in the room, this would heat the house. (In the case of the refrigerator, there is leakage through its walls to compensate for the heating of the room by the radiator.)

All this is possible because there is heat energy in practically all substances, but simply more of it in hot substances than in cold substances. Ice, for example, seems cold but it is warmer and contains more energy, for a given mass, than liquid air; liquid air is warm compared with liquid helium, which is near the absolute zero of temperature (minus 459° F or minus 273°C). Creating heat with a resistance heater takes as much energy as it would to raise its temperature or "pump it up" from near absolute zero. Pumping it up from temperatures that are only a few degrees lower than the space being heated takes much less energy, in proportion to the temperature difference. Running the electric motor of a heat pump thus makes full use of the high-grade electric energy.

The higher the temperature of the cool heat reservoir, the less electric energy is required to heat a house. An efficient scheme is to use a heat pump in conjunction with solar heating. Often the solar panels cannot quite reach the desired temperature but can warm a reservoir enough that rather little energy is needed for a heat pump to raise the temperature the rest of the way.

The electric power to heat a house, either with resistance or a heat pump, can come from the power line or from a windmill. This application is favorable for wind power because heat is a form of energy that can be stored economically, to be used later during windless periods, without requiring expensive storage batteries.

It is particularly advantageous to combine heating by wind power with direct solar heating employing the same storage medium because often the sun shines when the wind is not blowing and vice versa. Hence there is less dependence on storage. Even so, it may be uneconomical to provide a large enough storage capacity to be quite sure of riding out occasional periods of particularly unfavorable weather; it may be more economical to depend on commercial electric power as a backup. The combined scheme means less use of the backup. When commercial power is occasionally called upon, it may be arranged to do so only during the hours of minimum demand, to avoid adding to the peak load, since there is storage for more than a day. This would not require additional commercial generating capacity even if this scheme should be widely adopted.

The simplest and probably cheapest arrangement is to use the electric power from a wind dynamo for resistance heating of the same water tank or other storage unit as that heated by solar panels. This is not an ecological sin, as we have called resistance heating with commerical power, because one does not start with heat and there is no heating of cooling water at the power plant. The small ecological sin of occasional use of the backup power is more than atoned for by the power saved.

More elaborate variations of this scheme involving a heat pump require a greater initial investment. The heat pump could be powered most of the time by wind power, even when the wind is too gentle to provide additional heat to the reservoir, and some of the time by the backup power. There can be a second storage medium at a temperature higher than can be provided most efficiently by direct solar heating. Again, the heat pump could serve also for air conditioning, increasing the economic appeal of the system. Efforts are being made under federal auspices to perfect such arrangements and to encourage production to decrease the cost of components. A well-insulated small house demonstrating combined wind and direct solar heating, a part of this effort, is shown in figure 25. The 25-kilowatt wind dynamo is simple and rather graceful in appearance and not considered an eyesore.

Visual Acceptability of Home Windmills

Home windmills are probably practical only for homes that have some ground around them, such as separate homes in less densely populated parts of cities, suburbs, and beyond. Experimental small ones have been demonstrated on city apartment houses, however. Even in the suburbs, it may be that a windmill in almost every backyard might be considered unacceptable visual pollution. It surely would change the general appearance of things, but people already tolerate forests of TV antennas. Windmills are more conspicuous but might be considered more graceful.

Some years ago Merida, the capital city of Yucatan, near the famous Mayan ruins, was mainly a city of simple one-story houses with a Spanish influence, all very picturesque as if from a previous century. But if one looked over the flat city from a vantage point above house level it was a forest of metal windmills, each just like the water-pumping windmill of a midwestern farm (fig. 18) and imported from near Chicago. The reason for this is that the city is

Fig. 25. Demonstration residence heated by wind power and solar panels on the campus of the University of Massachusetts, Amherst. This modest 25-kilowatt wind dynamo was the second largest in the United States, the largest one in operation in early 1977.

built on a limestone layer under which is a copious aquifer, so there was no need for installation of a city water supply. It was a matter of every householder for himself, presumably first with a hand pump and then with a windmill. The windmills did not prevent Merida from being a very picturesque and attractive city. Perhaps windmills could become as acceptable in the United States, particularly the slender electric-power kind that are less conspicuous than the many-bladed water-pumping kind.

The "America the Beautiful Fund" recently circulated a group of pictures, each entitled "America the Beautiful," including Bryce Canyon, the Tetons, surf on a rocky coast, a covered bridge with fall foliage and, against a blue sky, a farm windmill.

4

Large Wind-Power Machines and Installations

While various small-scale solar-related sources for individual homes and business, together with conservation measures, can significantly reduce the drain on the commercial grid, this grid plays a key role in our industrialized civilization. Providing an adequate generating capacity remains a vital element in solving the overall power problem. Partly because of conservation and local sources, the historic exponential growth of energy output, with a doubling every ten years, will not be perpetuated in the future. It seems inevitable that the total power in the grid will ultimately level off if civilization continues to develop without cataclysmic setback. Presumably the rate of growth will be slowed, but not drastically cut off at this point. The question is not whether more large power plants will be built, but what kinds they will be. It is important to understand the extent to which they could be composed of large wind dynamos.

Anticipation of large-windmill development is justified by the experiences in Europe and the United States that have been described in chapter 1, particularly the experience with the 1.25-megawatt Smith-Putnam machine at Grandpa's Knob. The very fact that that machine, and several lesser machines in Europe, have been successfully phased into the commercial power grid provides a starting point for rapid expansion of large wind-power systems. It shows that the main engineering problems are already solved and further development is mainly a matter of refinement for the sake of improved efficiency and economy.

The Smith-Putnam effort included a study to determine the most economical size of windmill for this purpose. The results indicated that machines between about 1.25 and 2.5 megawatts should be most economical, but the curve of cost with size is quite flat across this range, varying by only about 2 percent. A similar British study quoted in Golding's book (1956) shows a curve that differs slightly from this; rather than rising slightly after 2.5 megawatts, it continues to slope down, suggesting greater economy at still larger sizes. A recent study by Lockheed finds the greatest economy at about 5 megawatts.

The size of the Grandpa's Knob machine was selected to be the smallest within what is considered to be the economical range. These estimates apply to the contemplated construction of modest numbers of windmills at a time and suggest that the early part of a large wind power program should be dominated by machines in the megawatt range, perhaps several megawatts. As the power production gets very large, there may be a trend to somewhat smaller machines in the 100-kilowatt range to save on construction materials, as is discussed in appendix 2.

As we speculate about the future trends, it seems to be a likely extrapolation to assume that the major machines will be of the rapidly rotating, horizontal-axis type, in this respect similar to the Smith-Putnam machine. It is possible that one of the vertical-axis types might win out on land after a period in which both types are deployed, but it seems more likely that any practical utility the vertical-axis types may have will be in the small-machine field. For the sake of making rapid progress toward the mass production of large machines, before energy options are frozen by pursuing only the fuel-consuming technologies, it seems important that national planning should proceed on this assumption and not await gradual growth of the vertical-axis types that might compete only after a period of development.

While the final preponderant size and shape of these machines may be expected to evolve with future experience, it is reasonable to project the possibilities on the basis of approximately 2-megawatt wind turbines, somewhat larger than the Grandpa's Knob machine. Right after the experience with that machine, two people in this country did some conceptual designing to indicate how they thought the next generation of machines might look. Palmer Putnam himself, working with the Smith Company engineers, designed a "pre-production" model and speculated on further cost-saving alterations of the Grandpa's Knob model. These studies suggest a machine quite similar to that one with the same blade span, 175 feet, and producing somewhat more power, 1.5 megawatts. The blades were to be twisted in four steps and power regulated by controllable wing flaps, placed just ahead of the leading edges of the blades, rather than by controlling the pitch of the blades.

The other conceptual design following the Smith-Putnam experience was that of Percy Thomas of the Federal Power Commission. As shown in figure 7, it called for two wind turbines totaling 6.5 megawatts mounted on a tower over 450 feet high, about three times the height and five times the power of the Smith-Putnam wind dynamo. The scale suggested by the house drawn under the tower in the figure illustrates how massive a structure might be needed to take advantage of the stronger winds aloft.

In the course of the recent revival of interest in wind power, it has been suggested that one should go even higher to reach the steadier and stronger winds aloft, perhaps almost a thousand feet with each tower carrying several wind turbines above the western Great Plains (Heronemus, 1972). Other studies have come to the conclusion that low towers carrying single turbines are probably more economical despite the less favorable winds at lower altitude. Based on this conclusion and on experience with the 100-kilowatt wind-electric generator near Stuttgart already mentioned (fig. 10), Professor Ulrich Hütter and his research institute for wind energy technology, (FWE at the University of Stuttgart), have proposed an interesting new "leaning tower" concept. In this design the weight aloft is minimized by transmitting the power through a drive shaft down the tower to the gears and electric generator at its base (Hütter, 1975). The tower is kept light by requiring only strength enough to withstand a storm when it is leaning into the wind (as a person would naturally do), as shown in figure 24. It is of course also strong enough to stand up in its leaning position when there is no wind. This design requires a very strong bearing at the base of the tower to permit the whole tower to swing around (or yaw) into the wind, rather than having just the assembly at the top of the tower do so as in the usual design. This turning into the wind is done automatically by a fantail, as was done with old-fashioned windmills in England of the eighteenth century (see fig. 2).

The leaning tower design could have great economic importance in large-scale deployment because it uses much less material than the usual design. In the engineering specifications for such a machine (considerably larger than the Smith-Putnam generator, but designed for weaker winds and a lower cut-in speed so that its power rating is slightly less), the weight per rated kilowatt is only 37 percent and the weight per unit of area swept by the blades only 14 percent of that of the Smith-Putnam machine. These figures and those of the earlier Stuttgart 100-kilowatt machine (fig. 10), also rated for a weak wind regime, are shown in table 1. The latter machine has its generator aloft but achieves relatively modest weight by using composite fiberglass blades and a pole braced with guy wires in place of the usual massive tower.

TABLE 1. Specifications of Three Wind Dynamos

	Height		*Blade Span*		*Area*	*Mass*	*Mass/Area*	*Mass/Power*
	m	*ft*	*m*	*ft*	*(m^2)*	*(1,000 kg)*	*(kg/m^2)*	*(kg/kw)*
Smith-Putnam, Vermont 1.25 Mw	42	140	53	175	2,200	210	95	170
Stuttgart 100 kw	21	75	34	111	900	13	14.3	130
Leaning Tower of Stuttgart 1.0 Mw	48	157	80	262	5,000	63	12.5	63

Siting of Large Wind Dynamos

The three favorable sites for large wind power installations are: (1) mountaintops and ridges; (2) the western Great Plains; (3) at sea offshore.

The Smith-Putnam enterprise chose a mountaintop. Meteorological towers were erected at several locations to observe wind-velocity profiles before Grandpa's Knob was selected as favorable. In hilly Vermont this is the natural choice. On a nationwide scale, and with the thought of deploying thousands of such machines in the future, it is unlikely that mountaintop locations will be the most advantageous. Very favorable winds are found there, but it is unlikely that a large enough number of favorable sites, both accessible and yet remote enough to overcome aesthetic objections, will be found.

The distribution of average winds throughout the contiguous United States and nearby offshore is shown in figure 26. There are three great concentrations of wind, two of them offshore, both just over the horizon from the New England Coast and the coast of the Pacific Northwest. The third is a large region of the western Great Plains extending into the eastern edge of the Rocky Mountains, with the most favorable winds in southern Wyoming, in the Oklahoma and Texas panhandles (Nelson, 1974), western Kansas, and reaching up through western Nebraska and the Dakotas. Most of that land is used for marginal grazing, with many acres per head of cattle. Some of it is used for growing grain in irrigated circles half a mile in diameter in a checkerboard array with large unused spaces at the corners. Giant windmills would need to interfere very little with either of these activities and there is room for hundreds of thousands of them. Therefore one option is to concentrate the windmills in those two regions of highest winds, southern Wyoming and the panhandles, but the economic question arises whether the saving in wind turbine cost in the regions of highest winds will offset the long transmission lines needed to reach the major centers of population. Costs discussed in chapter 9 indicate that wind

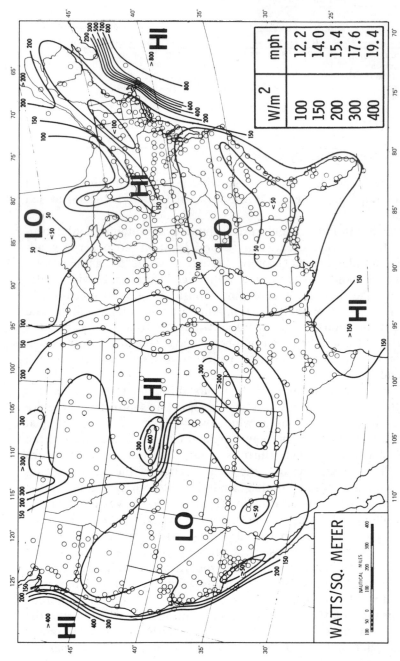

Fig. 26. Average available wind power. The individual stations (*small circles*) show fluctuations about the smooth average. The numbers indicated are in watts per square meter. (Reed, 1975).

W/m²		mph
100		12.2
150		14.0
200		15.4
300		17.6
400		19.4

WATTS/SQ. METER

power is economic even with such long transmission lines. If super-conducting power transmission should become practical in the future, this could make it more so. In any case, a substantial start can be made by supplying the electric power for the populations and industries near those regions. Five-hundred-mile transmission is considered practical (with a power loss of about 5 percent) and would reach as far as St. Louis and Houston from the panhandles–western-Kansas region.

The offshore winds, particularly those off the northern half of the eastern seaboard, are even more favorable. It may be noted in figure 26 that the average winds within a hundred miles of the tip of Cape Cod attain a value of 700 watts per square meter whereas those in Wyoming reach a bit over 400 watts per square meter. The area of such high offshore winds is comparable in size to the high-wind area of southern Wyoming, with plenty of space for tens of thousands of huge floating windmills. A further advantage of this site for windmills is that it is close to the densely populated seaboard energy market.

The question of visual pollution and compatibility with other space uses is quite different at sea. A very large array of floating windmills half a mile or so apart would clearly present a hazard for navigation and they would have to be arranged to leave ample navigation lanes.

The favored location for the first several thousand offshore windmills is George's Bank, just north of Cape Cod. This region was formerly the site of thousands of commercial fishing boats and shipping lanes avoid it; the main transatlantic lane out of New York passes to the south of it. The main concern, then, is possible interference with the interests of the few remaining commercial fishermen. They will of course have to take care to avoid colliding with the windmills, as they do now with each other, but there may be a compensating benefit. A former commercial fisherman, when asked about this, has pointed out that a boat sometimes finds a favorable fishing spot with a large school of fish and wants to return to the same spot after unloading its catch in port. The navigation used is not always accurate and a grid of convenient navigational fixes might be very welcome. The windmills would thus double as spar buoys. There is already one Texas tower on the southern edge of George's Bank that has provided useful information about wind and waves there.

Offshore windmills may be either floating structures or structures similar to land-based windmills, rigidly mounted on the sea bottom. The floating type, each of course firmly moored to a heavy anchor weight, is favored at depths of several hundred feet. The depth must be great enough that the floating structure will not contact the bottom even when dipping in the troughs of the enormous waves (nearly fifty feet high) of the most severe storms. Rigidly mounted structures would be favored for shallower regions that are usually near shore, but there is less prospect of deploying large numbers of these because of expected aesthetic objections and because the winds are not as strong as farther offshore (see fig. 26).

Large Wind-Electric Arrays on Land

The vast expanses of the windy western Great Plains present the opportunity for large arrays of electricity-generating windmills of various types and sizes. It seems likely that they will eventually evolve to a preponderant type and size proven by experience in that environment to be the most economical and effective. However, it would be a mistake to let large-array deployment await the results of small-array evolution, or to confine the large-scale deployment to one type and size, without simultaneously exploring some of the others in considerable numbers. If the choice of sizes and types is sensibly made, there will probably not be much difference in performance and all the wind dynamos built will probably be useful as long as they last, resulting in a variety of machines feeding the wind-electric component of the grid as is now the case with fuel-powered grids.

The two principal types to be considered are the horizontal axis and vertical axis. As we have remarked, the conventional horizontal-axis type has been proven successful and should receive the predominant emphasis in planning initial large-scale deployment, unless the other type achieves outstanding success soon.

While the Smith-Putnam and other studies of horizontal-axis machines on hills or mountains have concluded that the most economical sizes are in the range above 1 megawatt, this conclusion should not be accepted as final for large arrays of windmills in flat country where the area of suitable land is not a limiting factor. In these earlier studies, the unit-power cost of windmill installation goes up when output is less than 1 megawatt. Two of the reasons for this rise are (1) the greater cost of the detailed work to make a large number of smaller machines and (2) the siting costs may be almost the same for a small or large machine when it is necessary, for example, to build a road up a mountain to a site. The first of these disadvantages might be greatly reduced in true quantity production and the second in flat country where siting is simpler.

It may thus not be clear at the start whether, for example, an array of a hundred 2-megawatt windmills or of a thousand 200-

kilowatt ones would be better. Probably both should be built initially while production facilities are being organized for producing larger arrays.

It might turn out that one is more profitable financially but that the other better conserves critical materials. In this case, the decision should probably be made on the basis of national priorities rather than straight commercial competition, perhaps with government subsidy arranged to tip the balance. The questions of materials economy and energy payback time will become increasingly important as very large systems progress.

As long as the windmills are of the same general design, the production of a certain amount of power by a larger number of smaller windmills is expected to require less material but to occupy more land. If the smaller windmills are one-tenth as powerful as the larger, as in the example just cited, a simplified calculation explained in appendix 2 estimates that about one-third as much material would be needed and the array would extend over about twice as much land.

The land area of the array is determined by the spacing between adjacent windmills. If the spacing is ten times the blade span, for example, to avoid a serious sheltering effect, the smaller windmills will be closer together than the larger ones, but not enough so as to compensate for their greater numbers, therefore their array will be larger.

Thus the determination of the "best" size for the windmills, if there is one, will involve a balance between several factors: the amount of construction material, land area, unit construction cost, and cost of maintenance. Since only the first of these four clearly favors the smaller machines, it seems likely that the balance will be slightly in favor of the larger windmills of perhaps 2 megawatts or so, at least until arrays become large enough to make serious inroads on materials availability.

The land-area requirement could turn out to be less important than the materials requirement because there is so much land not under intensive use in the windy regions. A widely spaced array of windmills would not interfere seriously with present use of the land, whether used for grazing or agriculture. The bases of the windmills themselves would occupy only a very small fraction of the land.

Insofar as the flatness of the land permits, an array will presumably be of a regular pattern, perhaps a giant checkerboard array with windmills at the corners of the squares. The spacing or lattice distance will be determined by the length of the "wind shadow" cast by each windmill, with due recognition of the fact that the wind seldom blows exactly along the direction from one windmill to the next. One preliminary study (Templin, 1974) that involves a two-dimensional model, apparently not recognizing this fact, suggests that a spacing thirty times the blade span may be required. This would mean placing machines like the Grandpa's Knob wind dynamo a mile apart. Intuitively this seems too far and indeed a 1945 study at New York University favored spacing at ten times the blade span. Further studies are needed and would be desirable if two or more of the early large demonstration wind dynamos were built rather close together in line with the prevailing wind. This would allow observation of the sheltering effect in the natural meteorological flow, an effect that cannot be duplicated in a wind-tunnel experiment. Pending such investigations, we shall assume for the sake of rough orientation that a spacing ten times blade span will be enough to prevent serious interference between windmills.

There is enough area in the windy parts of the western Great Plains that space need not be a limitation on the exploitation of wind power for a long time to come. According to the rough indication of figure 26, the windiest region of the Great Plains is in the Texas and Oklahoma panhandles and in western Kansas. In that figure the elliptical contour line for 300 watts per square meter encloses a windy area of about 20,000 square miles. If this were covered with an array of 1-megawatt windmills of 165-foot blade span ten times that far apart, thus with ten windmills per square mile, the array would have a rated power of 200 thousand megawatts, or 2×10^5 Mw. That is as much as the rated power of about 200 large (1,000-megawatt) nuclear power plants. Of course neither system produces at its rated power. For large nuclear power plants the performance figure is about 55 percent. In that windy region windmills might do almost as well, but it is safer to assume an average performance of about 40 percent of their rated power. This means that the average power of the large windmill array would be equal to that of about 40/55 of 200, or about 145 large nuclear power plants. These figures are just for the windiest region of the western Great Plains. The area having a wind intensity of over 200 watts per square meter is about ten times as large. If the array were extended to cover this great area, it would supply more than the nation's energy needs for the foreseeable future.

These figures emphasize that the difficulty in meeting a major part of our energy requirements is not land availability, but rather the task of building and financing an enormous number of large wind machines. When contemplating this formidable prospect, one must judge it in relation to the great accomplishments already achieved by our industrial capacity.

The Storage Problem with Large Land Arrays

The fact that the wind does not blow all the time requires special planning for the use of power from windmills. This problem becomes more acute as one contemplates very large wind-power systems supplying a large fraction of the nation's total electric power. It is also true of present fuel-consuming power plants that they do not perform all the time, but the frequency of power interruptions is somewhat different from that expected with wind power. They typically perform day in and day out for some weeks or months, and then require shutting down for some time, whereas the wind is apt to fail frequently and briefly day to day and even hour to hour locally. In either case it is necessary to have a reserve capacity of some sort, either a backup generating capacity or energy storage to supply the electric gird. This extra capital investment would not be needed if the sources performed continually. In the case of fossilfuel-consuming plants, the plants using the cheapest fuel are normally used most and the backup plants may be those using more expensive fuel. In a combination of fossil-fuel and nuclear plants, the nuclear plants carry the base load with the fossil-fuel plants as backup, because nuclear plants deteriorate if turned on and off frequently and because their expense involves capital investment more than direct fuel costs.

In a grid supplied mostly by wind power, with the windmill arrays in widely separated regions, the hour-to-hour fluctuations will be smoothed out and fossil-fuel plants can act as backup for the day-to-day fluctuations. But this is an expensive arrangement, because the wind component does not directly save capital expenditure in the fossil-fuel component and would mainly conserve fuel. With recent high fuel prices, even this arrangement is estimated to be economic, with the fuel saving compensating the wind-power investment (Coty, 1977), as is discussed in chapter 9. Fuel conservation is of course important ecologically as well.

The additional wind-power component does save capital expenditures indirectly because the determination of how much backup is needed in any system is a statistical matter. Even with a lot of backup as insurance against breakdown of some components, it can happen that all components break down at once, as occurred in Florida in May, 1977, when the Miami area was blacked out for a time. The purpose of the backup is to minimize such an occurence.

With more diverse components of the system there is less chance of total failure; in this sense each component of the system backs up the other. The wind component has its place in the mutual backup and reduces the need for reserve capacity in other components of the system.

As in the case of other generating units, an individual wind dynamo will occasionally have to be taken out of service for repair. This is one reason to have backup, but the smaller the individual generating units and the greater their number, the better. An outage of a 2-megawatt wind dynamo, or of several of them, is a much less serious disturbance of the system than an outage of a 1,000-megawatt fuel-consuming plant.

Thus, in conjunction with other types of power generation, it is possible to use quite large arrays of wind dynamos to feed the electrical grid without special energy-storage facilities. However, storage facilities would contribute significantly to the mutual backup arrangement. Storage, to smooth out the day-to-day wind fluctuations, becomes essential if wind power is to carry the major part of the utility electric load. Large amounts of electric power supplied by the commercial grids are used for industries that do not depend on a continuous power supply, such as aluminum and fertilizer production. These industries could be supplied almost as well by wind power without storage facilities, but with increased capital costs due to equipment idleness. Such industries have already sprung up where hydroelectric power is available and could expand into windy regions where wind power can be harnessed most economically.

For the initial stages of large-scale wind power there is storage available requiring no new technology and practically no extra cost —namely, hydroelectric generating capacity. Its output may be increased by the use of wind power, as was contemplated in connection with the Grandpa's Knob experience in Vermont years ago. The output of most of our great hydroelectric installations is dependent on river flow into a reservoir, the installed generating capacity being more than enough to take care of the year-round flow. The generators are easily turned on and off to meet varying demand. If the hydroelectric installation teamed up with wind power, the wind would carry the load to its maximum capability while the wind is blowing, permitting the river to go on filling up the reservoir, and the hydroelectric generators would draw on the replenished reservoir to carry the load when the wind dies down. The reservoir thus serves as backup storage for both the hydroelectric power and the wind power. It has been estimated that there is enough extra storage capacity behind dams in the Pacific Northwest that wind power, used in this way, could increase the available electric power there by more than 50 percent.

There is one method of energy storage now in common use in connection with fuel-consuming power plants known as pumped hydroelectric storage (Warshay, 1976; Newark, 1976). A typical installation is shown in figure 27. It is very different from the existing hydroelectric facilities just mentioned in that water is first pumped into an upper reservoir by power from the steam generating plant before it is used to drive the generators when more power is needed. Here some inefficiency is encountered; only about 70 percent as much energy is generated as is used for the pumping. The method is used for smoothing out daily fluctuations in demand, providing for the peak demand that usually occurs in the early evening without the need for additional steam or diesel generating capacity. During the slack period late at night, existing generating capacity is used more fully.

While the use of this method has until now been limited to hilly or mountainous terrain where a high reservoir could be established, it is expected that underground reservoirs will remove this limitation (Newark, 1976). This is an established storage technology available for use with wind power, smoothing out the unsteadiness both of the wind and the demand for power.

As is mentioned in chapter 9 where costs are discussed, there are hopeful prospects for other methods of storing energy that are capable of large-scale expansion, but these require further research and development. Probably the most hopeful of these is electrolytic conversion of wind-generated electric power to hydrogen gas. The gas may be stored directly or converted into a more easily storable chemical compound and then converted back to electric power by use of fuel cells or other means. The advantage is that the actual storage of the gas underground or deep underwater would be relatively cheap, so storage for several months would be practical. The disadvantage is that the cycle is inefficient; about half of the input energy is lost in the process. Another possibility being investigated is flywheel storage based on the availability of new strong materials (Post, 1973) but this is promising only for daily peak smoothing and not for the several-day storage needed for wind power (fig. 28).

These storage methods require further research and will only become economically feasible with the development of mass production. This should fit well into a program of building up industry's capacity to produce giant wind dynamos initially for moderately large arrays that would supplement hydroelectric capacity and other sources. Ultimately, as the production rate increases, the dy-

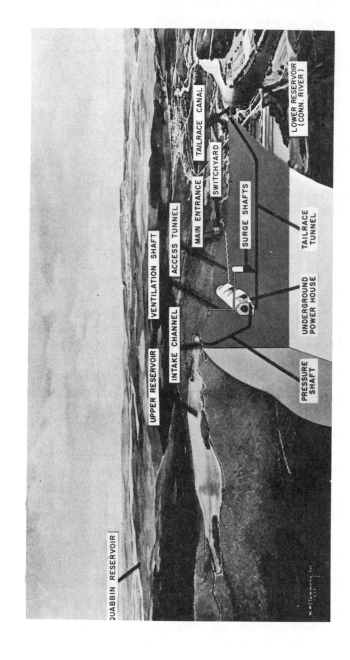

Fig. 27. Pumped storage facility at Northfield, Massachusetts. (*Courtesy Western Massachusetts Electric Company.*)

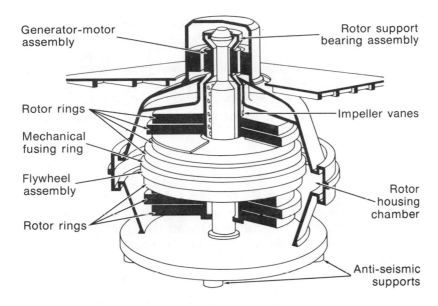

Generator-motor assembly

Rotor support bearing assembly

Rotor rings

Mechanical fusing ring

Flywheel assembly

Rotor rings

Impeller vanes

Rotor housing chamber

Anti-seismic supports

Fig. 28. Flywheel storage unit proposed for meeting daily peak demand. The fiber-composite flywheel would be about 15 feet in diameter and weigh about 200 tons. The chamber is evacuated. (Adopted from ''Flywheels'' by Richard F. Post and Stephen F. Post. Copyright 1973 by Scientific American, Inc. All rights reserved.)

namos would supply a major part of the power in conjunction with newly developed pumped hydroelectric storage and future methods.

Offshore Floating Wind Dynamos

The idea of a large array of large floating windmills offshore above the continental shelf is an exciting prospect. Unfortunately, there has not yet been any practical experience with this aspect of wind-power technology. It obviously poses new engineering challenges but should nevertheless be a straightforward development applying modern technical capabilities. It may even be considered a low-technology undertaking in the sense that the principal part of the needed development is almost routine mechanical engineering and nautical design, requiring none of the sophisticated complexity of, for example, a nuclear reactor. This does not mean that it will not have its problems, especially vibration analysis, that is encountered in any large wind turbine, and the rather special gyroscopic effects associated with an unsteady base. These should be solvable by straightforward good engineering. The operation can also benefit from modern electronic controls involving simple computers, but this would be an incidental part of the massive undertaking.

Most of the ideas about offshore wind power stem from the pioneering proposals of Professor William Heronemus, a civil engineer and also a navy captain with broad nautical experience not only at sea, but also in submarine and other naval construction and design.

One might first think of a floating windmill as one mounted on a broad platform or barge at sea, one wide enough so as not to capsize. Instead, one proposed design (fig. 29) attains stability by having ballast beneath its flotation, as does a ship, but with one large hollow sphere as flotation and some distance beneath it, a smaller heavy sphere as ballast (Heronemus, 1972). A sphere is the most economical shape to enclose a volume. The flotation sphere does not float on the surface but some distance beneath it, since some flotation is provided by hollow vertical spars that form part of the tower supporting the superstructure. The flotation of these spars determines the depth at which the structure rides and only these spars are awash in the surface waves. This arrangement greatly reduces the force with which the waves buffet the structure, as compared with having the main flotation hull on the surface. To some extent, it mimics the way a submarine escapes the violence of the waves.

Such a flotation unit may be used to support a single wind turbine or several of them. It seems that two or more may be preferable to a single one. There are advantages in floating a large structure, just as there are advantages in floating a large ship (large compared with the size of the waves, so the perturbations caused by the waves may be relatively small). Yet there are limits to the practical size of a wind turbine; a large flotation unit may better carry several turbines of moderate size than one very large one. Thus, the tentative conclusion that single-turbine units are best on land need not hold at sea.

Heronemus has proposed and made preliminary stress calculations on a three-turbine unit, each turbine generating 2 megawatts for a total of 6 megawatts, with the turbines supported about three hundred feet above the surface on an aluminum frame, as shown in figures 29 and 30. This is intended for offshore deployment where the winds are high, where the relatively high cut-in speed characteristic of large wind turbines is not a great disadvantage.

When production of wind dynamos is well along, there will probably be economies in building them in large, multimega-

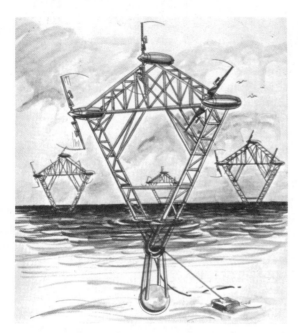

Fig. 29. Heronemus's conceptual design of a 6-megawatt offshore unit. (Heronemus, 1972).

watt sizes that may offset their lower cut-in speeds in windy regions. Initially, however, the use of multiple arrays of smaller wind turbines may be better, because the smaller generating units will sooner achieve the economies of mass production, as well as having lower cut-in speeds. A design of this sort is shown in figure 31. It employs a floating platform consisting of many vertical cylindrical floats, so as to present less area to storm waves than would a solid hull. Its individual wind turbines are the same model having thirty-five-foot blade circle that has already been tested, along with its automatic controls, in the home-heating installation shown in figure 25. There, in a moderate wind region, the generator capacity is only 25 kilowatts, but in the stronger winds off the New England Coast, the rating would be 60 kilowatts and in the still stronger winds in some parts of the high seas, 100 kilowatts. At 60 kilowatts each, the seventy wind dynamos of figure 31 would yield about 4 megawatts, or two-thirds as much as the spar-buoy design of figure 29. If it should turn out that floating units are also economic in places like the Great Lakes where the winds are more moderate, such a multiturbine design, with lower power ratings and lower cut-in speeds, would be appropriate.

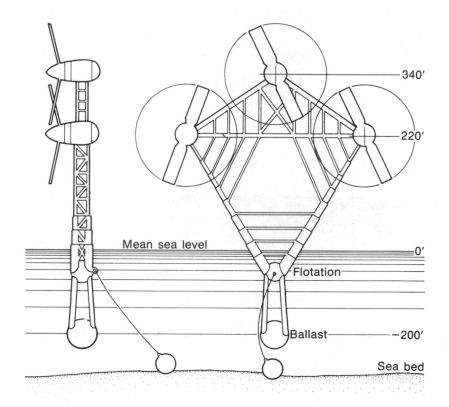

Mean sea level

Flotation

Ballast

Sea bed

340'

220'

0'

−200'

Ready for towing to site

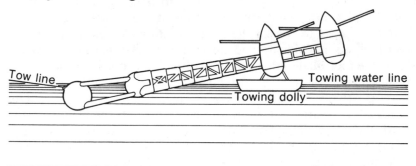

Tow line

Towing water line

Towing dolly

Fig. 30. A floating wind dynamo, such as this Heronemus design, would be assembled in a shipyard and towed to its mooring. (Heronemus, 1972.)

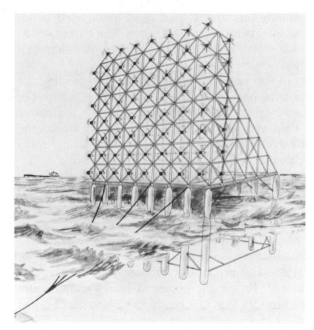

Fig. 31. Heronemus's offshore platform model. (*Courtesy Prof. Heronemus.*)

The storms are severe in the windy offshore regions. Waves over a hundred feet high have been observed at the Texas Tower off Nantucket. There is some difference of opinion as to whether a well-designed wind turbine unit would survive the storm and icing conditions there. For this reason it is regrettable that we do not already have one or more floating wind dynamos there, testing and demonstrating this potentially very important power source. One cannot reasonably make further plans without such experience. This represents perhaps the most serious shortcoming of the federal wind-power demonstration program.

Any attempt to imagine the nature of a large array of offshore floating windmills before the first one has been built and demonstrated must be considered tentative and speculative. Nevertheless, it is useful to exercise some imagination and to outline what a large array might be like in order to set up tentative goals indicating the realistic sizes of such arrays. The prospect of taking a first big step soon toward such a goal should provide a vigorous incentive to develop the first few floating units. Without such a vision of the future one could hardly get started, since the power produced by the first few units cannot be expected to pay for the development costs or to motivate and justify the initiative.

The general considerations about spacing of land-based wind-mills to avoid shielding effects also apply offshore. The higher winds offshore allow somewhat closer spacing, although not closer than half a mile for multimegawatt units. Even with half-mile spacing or more there is plenty of room for very powerful arrays. The continental shelf area off the northern half of the eastern seaboard, where the winds are over 300 and up to 700 or 800 watts per square meter (fig. 26), is greater than the area in the Great Plains, where the winds are shown to be over 300 watts per square meter. Until arrays begin to contain a hundred thousand multimegawatt-scale wind-mills, one need not be concerned about there being enough space, offshore or on the plains.

The problem of getting the power to where it will be used is somewhat different offshore than on land. The principal options are: (1) to transmit the power by submarine cable to shore and feed it into the established electric power grid; (2) to use the power lo-cally to electrolyze water and produce hydrogen to be transported ashore by pipeline; and (3) to use the power at sea to manufacture some power-intensive product that can be shipped ashore. This last possibility is contemplated for ocean-thermal-difference power plants floating in the Caribbean too far from shore for direct power trans-mission. The plan, which is discussed in the following chapter, is to ship ammonia used in making fertilizer. Since offshore windmills will be much closer to industry and populations, transmission of the power ashore, either by electrical cable or by hydrogen pipeline, is to be preferred.

Submarine cable transmission of electric power is more efficient with direct current than with alternating current, because of the electrical property known as distributed capacitance of the line. As an example of this practice, power is transmitted from mainland Sweden to some of its islands in the Baltic by direct current submarine cable. One attractive possibility for an array of offshore windmills would be to transmit the power directly ashore in this way, where it could be both used industrially as direct current, and transformed to alternating current and fed into the power grid. This would be the simplest use of the power, if not the most economic, and is preferred for the first demonstration windmills and the first offshore arrays that need not be very far from land. For the first offshore demonstration, a site twenty miles east of Chatham, which is at the elbow of Cape Cod, would be ideal. The site offers a satisfactory depth and good wind, yet is near enough to be connected to land by a rather short cable. A region of moderate depth known as George's Bank extends almost a hundred miles to the east and north of that site and is also a favorable site for a large array that could be connected to shore by cable.

Transportation of energy in the form of gas by pipeline is much cheaper than in the form of electric power by transmission line. Yet on land it is not the practice to convert electric power to gas for transmission over long distances and then convert it back into electricity, presumably because of cost and inefficiencies of the conversion process. Simplicity and convenience direct that the energy arrive over the high lines in the form of electricity.

For large offshore wind power arrays, the hydrogen-transmission option looks particularly attractive because of the possibility of combining it with a promising solution of the storage problem. This would establish a self-sufficient power system. On land, large strong steel tanks for storage of hydrogen gas under high pressure would be prohibitively expensive, but at sea the pressure of the deep ocean can be exploited to help contain the high-pressure gas. Large collapsible containers, inflated by the high-pressure gas deep under water, would be used rather than strong, rigid steel tanks.

In addition to the floating windmills, such a system would require electrolyzer stations, a pumping station, and collapsible storage vessels, all at sea. These stations would probably be subsurface hulls, like submarine hulls moored at a suitable depth to avoid severe wave action. The system would have the capability to deliver hydrogen at a continuous rate by pipeline to shore. There the gas would either be used directly as a continuous fuel supply in industrial processes and for heating purposes, or else part (or all of it) could be converted to electric energy for the commercial power grid.

With present technology the conversion from hydrogen gas to electric power would take place by way of a steam cycle, just as is done now with natural gas. In common with all steam-electric plants, it is an inefficient process because it involves the Carnot loss of heat to the condenser coolant. Future technology will probably include large-scale economic use of fuel cells that are now used only on a small scale. This technology is being developed and deserves vigorous encouragement. It will probably be practical within a decade or so. It's a reasonable hope that by that time offshore wind power would be developed through the stage of modest-sized arrays feeding electric power ashore by cable. When the production facilities have reached the stage of manufacturing large-scale offshore arrays, they could thus use hydrogen storage and transmission, the hydrogen being used for fuel-cell conversion to electricity ashore, as well as for making easily transportable fuels and for heating by direct burning.

In suggesting these ideas, Heronemus describes and estimates the cost of one sample of the type of large array which we could be rapidly developing now (Heronemus, 1972). His specific example would consist of 13,600 of the 6-megawatt wind-generator units shown in figures 29 and 30. Each group, about 170 units, feeds its electric power into a submerged electrolyzer station. Each of the eighty-two electrolyzer stations feeds part of its hydrogen by pipeline into a single pumping station, submerged in water of moderate depth near the edge of the continental shelf. This station pressurizes the hydrogen for storage in collapsible, deep-water storage vessels, just over the continental shelf where the pressure is great enough that hydrogen can be stored in containers of moderate size without depending on their strength. Part of the hydrogen goes ashore directly by pipeline without pressurization, the rest is stored and later goes by the same pipeline after depressurization at the pumping station.

The cost estimate includes a preliminary estimate for fuel-cell stations ashore to convert the power to electricity. The rated generating capacity of the entire system is 82 thousand megawatts and the cost projection made by Heronemus in 1972 comes to $22 billion. That would be about $27 billion in 1977 dollars, or about $330 per installed kilowatt, storage and delivery to a network ashore included. This figure must be modified to take into account that the wind does not blow all the time. A plant capacity figure of about 33 percent to recognize that fact and allow for occasional repair time is reasonable, and would bring the capital cost for steadily delivered power to $1,000 (in 1977 dollars) per kilowatt. If this figure sounds high compared with other power sources, one must remember that they also do not perform all the time and, in some cases, one must multiply installed-kilowatt costs by a factor as large as two to get average-delivered kilowatt costs. Even acknowledging that this wind power estimate is optimistic, being based on projection rather than complete experience, it shows that offshore wind power seems to have promising prospects economically, as is discussed further in chapter 9.

Advantages of Offshore Siting of Wind Dynamos

The strong winds off the Atlantic coast and the proximity to the industrialized seaboard are the most obvious advantages of offshore wind dynamos. It is clear that some additional flotation components and techniques are needed to moor windmills offshore and it might seem reasonable to postpone this development until land-based wind power is progressing well. There are compensating advantages, however, that make it desirable to promote offshore wind power installations simultaneously with those on land, perhaps even with added emphasis offshore because we need to make up for the lack of experience there.

The strength requirements in the design of a windmill must not only withstand ordinary operating winds, but must stand up to a "killer wind," the most severe storm condition that might be encountered in a century or so. This design problem differs on land and at sea. The first and perhaps greatest difference is that the land-based tower must stand rigidly against the wind whereas the floating structure yields, softening the impact of gusts and leaning with the wind to reduce strains. The floating structure of figure 29, with its similarity to a spar buoy, is particularly effective. It does not have to be strong enough to stand up to the main force of the killer wind; it tilts, instead, reducing the bending movements. In its inclined position, it still has to be able to withstand the additional force of the superposed gusts without having time to react to them because of the inertia of the flotation members. Both because of the reclining position and because in such a storm the superposed gusts or fluctuations are not as strong as the main wind, this requirement is expected to be considerably less demanding than that of standing up rigidly against the main wind.

At sea, the strength requirement for the spars is complicated by the accelerations and direct forces due to waves. In normal weather and usual storms the accelerations of the base structure are minimized because the flotation members are mainly well beneath the surface level of the wave troughs so the members are subjected to only a regular circular motion within the wave, of diminished amplitude because of the depth. It is anticipated that this advantage can be retained in worst-storm conditions but this question requires further analysis in the course of designing details of the structure. During the worst storms the turbines will have been stopped to avoid gyroscopic effects.

The manner of assembling offshore windmill structures will also be quite different from that on land. Building an offshore windmill has some similarity to shipbuilding and will be done in shipyards or similar shoplike conditions, although important components are made elsewhere. This is better than construction in the field at each remote site. When completed and launched, as is a ship, a structure with the buoy type of flotation can be ballasted to float in a nearly horizontal position, as shown in figure 30, and towed to its mooring site. There water would be let into the lower sphere to make the windmill flip up.

Industrial Effort

The power of the wind is diffuse and very large systems are required to harness commercially important amounts of it. Building enough large wind dynamos to supply a major fraction of our electric energy needs would require an industrial effort comparable to some of our major enterprises, but electric energy is so vital that such a large effort would not be unreasonable.

A rough idea of the magnitude involved is conveyed by the following statement: The construction of enough megawatt-scale wind dynamos to supply the present electric energy consumption of the United States would require about as much metal and other materials as is used in three years' production of the automotive industry. Since both large wind dynamos and automobiles are machines requiring careful engineering and construction, the material consumed may be taken as a very rough measure of the industrial effort involved. On this basis one might expect that a production capacity about one-third as large as the automotive industry could in ten years produce wind dynamos to equal our present electric power generating capacity. While this does represent a massive industrial effort, it would not require building the whole industry from scratch since much of the work could be done by present industries, some of them expanding facilities to accommodate welcome additional business. A less ambitious goal would of course require a correspondingly smaller industrial expansion.

This estimate is made as follows: From table 1 we may take the weight of large wind dynamos to be roughly 130 kilograms per kilowatt or 140 tons per installed megawatt (though it might be considerably less in future machines). With a capacity factor of 40 percent to take into account the unsteadiness of the wind, this means that a wind dynamo weighs about 350 tons per average megawatt generated. The average electric power consumption in the United States is about 300 thousand megawatts (the installed generating capacity being about 400 thousand megawatts). Supplying this at 350 tons per megawatt would require wind dynamos weighing a total of about 100 million tons. The automotive industry in 1973 produced 10 million autos averaging about 2 tons apiece and 3 million trucks average 4 tons. The total production is thus about 32 million tons per year. Three years' production then weighs about as much as the required 100 million tons of wind dynamos.

A new industry of this magnitude, or one even a tenth its size, would not only be a stimulus to industry but also a boon to labor, reducing unemployment by providing a large number of jobs for semi-skilled as well as skilled labor. Just because of the diffuse nature of the wind, wind power is a labor-intensive source of energy. Such an industry could be built up rapidly with relatively little retraining of workers because it would include skills and facilities already used in other industries. The gears, hubs, and shafting require auto-makers' and heavy-machinery skills, the generators are a product of the electric industry, the blades have some similarity to airplane wings and the towers are akin to bridges and sky-scraper frames. Besides these, if offshore windpower is developed the shipbuilding industry would be extended to provide the flotation members. In some cases the growing wind dynamo industry would replace anticipated loss of expanding demand in other sectors as the looming ecological limitations of our finite resources dictate conservation and reduction of growth curves. The trend could be, for example, that workers now making big automobiles make big wind dynamos and smaller cars instead.

Back in the mid-thirties wind power on American farms was pushed out of the picture by the national rural electrification project that was motivated by a need to relieve unemployment during the Great Depression. Then fossil fuel was plentiful and cheap. Now wind power could be brought back into the picture in a much larger role to save scarce fuel and incidentally, once again as a measure to relieve unemployment.

5

Other Solar-Related Energy Sources

While wind power presents the most immediate prospect for large-scale generation of electricity, other solar-related sources are increasingly helping to meet world energy needs on a small scale now and should make important large-scale contributions later on. Among the direct uses of solar energy, home heating and cooling can now reduce the consumption of fuels, large solar steam-electric systems are entering the pilot-plant stage after very limited past experience, and solar cells are in need of development and drastic cost reductions before contributing to the national need for electric power. Among other solar-related sources, ocean-thermal generation of electricity is almost as ready for expolitation as is wind power, but on the basis of less experience, and some methods of bioconversion are already in use on a small scale.

Direct Solar Energy—Home Heating and Cooling

About a quarter of the energy consumption of the United States is devoted to space heating and to supplying hot water. About three-fifths of this is for residences and two-fifths commercial. These uses of energy do not involve very high temperatures making it relatively easy for solar energy to supply a considerable part of the need. Another 5 percent of our energy use provides air conditioning and refrigeration. These now depend almost entirely on electric power. In some parts of the country the air conditioning load creates the peak power demand on hot summer days when solar energy is most plentiful. This peak demand ultimately determines the number of electric generating plants that are built; thus, important savings in fuel as well as in demand on electric power plants could be achieved if a substantial fraction of the heating and cooling load could be carried by direct use of solar energy.

The simplest schemes for home space heating or hot-water heating in modern homes involve the use of solar panels. These are efficient absorbers of the heat of sunshine, transferring the heat to a circulating fluid, either water combined with antifreeze solution in northern climates or air. Each panel has the form of a shallow rectangular box, usually about four by eight feet in area and four inches thick. Several of them may be placed on the roof of a house, preferably on a south slope, to supply most of the heating needs. They can also be incorporated in the south walls of a house or placed flat on the ground using the south walls or a solid fence as reflectors to intensify the radiation on the panels.

A solar panel is a sort of sandwich (fig. 32). The central layer, the most important part, is a sheet of metal with a blackened heat-absorbing surface in good thermal contact (provided by soldering for example) with tubes to carry the circulating fluid that transports the heat to a storage medium. The bottom layer of the sandwich is thermal insulation that prevents heat loss from the bottom of the panel. The top layer is transparent glass or plastic, to admit visible light and impede the exit of heat rays. It thus traps the solar radiation. This is the well-known "greenhouse effect." It also excludes cold air in cold weather.

The other major component of the solar heating system is a

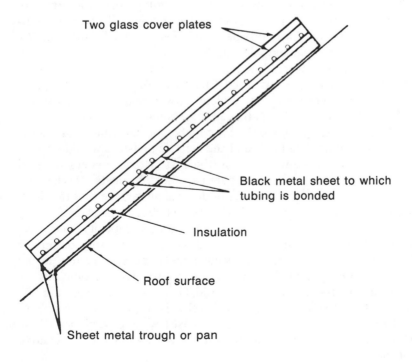

Two glass cover plates

Black metal sheet to which tubing is bonded

Insulation

Roof surface

Sheet metal trough or pan

Fig. 32. A roof-top solar panel for circulating water.

storage unit, which may consist of a water tank or a volume of rocks, usually in the basement or in the ground near the basement. From that point the heat is conducted through the house by ducts similar to those used with a conventional furnace, either forced hot air or hot water. When a solar source is added to a house with a furnace, both sources may use the same ducts, with the furnace reserved for use in long cold and sunless periods for which the storage system may be inadequate.

A major part of the cost of the solar heating system is the storage unit and if it were designed to carry over the longest cold sunless periods the cost would be prohibitive in most cases. The economic compromise is to combine the solar heating system with an auxiliary fuel-dependent source of heat. The avowed function of the solar heating system is then simply to save fuel and thereby to save long-term heating costs. If the system is installed during construction of a new house, there is a larger initial capital investment to be repaid by lower operating costs.

Addition of solar heating to existing homes is especially favorable in those with electric resistor heating. Many home owners find themselves stuck with this wasteful form of heating because many builders installed it on account of its low initial cost in a period of low electric rates. The prospect of higher heating bills was not serious enough to scare away prospective buyers. Indeed, until recently electric utility companies paid builders a bonus for installing it, unbeknownst to the buyer. Now that electric power rates are skyrocketing, both because of increased fuel costs and because of apparently unwise investments in nuclear power plants, home owners saddled with very high heating costs should find it attractive to install solar heating. It would be capable of carrying about three quarters of the heating load, and the electric resistors would be kept as the auxiliary system to supply the rest. Home owners with oil furnaces are also faced with increased heating costs, though not quite as great, and they have the additional advantage that the ducting for use with the solar heating system is already in place.

Even with a completely adequate storage capacity, there remains a need for electric power to drive the circulating pumps or fans, both in the loop between solar panels and storage and in the forced hot air system throughout the house. This fortunately requires much less energy than would be required to generate the heat. With solar panels at ground level it is in principle possible to avoid the use of such pumps but the cost and inconvenience of the larger ducts of a gravity circulation system leaves this alternative unattractive.

Simple heat panels collect heat for storage most efficiently at rather modest temperatures, storing warmth appropriate for house heating rather than high-temperature heat. Functioning on less sunny days, they are even more efficient if the warmth required is not quite sufficient for comfortable house heating. For this reason it is effective to use a heat pump between the storage unit and the house circulation to increase the temperature of the circulated heat slightly, using a correspondingly small amount of electric power as was discussed in chapter 3. During the long cold and cloudy spells when the storage is inadequate to keep this temperature difference small, the heat pump functioning as the auxiliary system has to work harder but still uses far less energy than would resistance heating.

The heat pump alone, without the solar energy system, provides a method of house heating that is much more efficient than resistance heating, using much less electric power though it has a higher initial cost. In this case, it has to increase the temperature from the cold outdoor temperature or a cool water source up to the temperature of the warm indoors, a considerably greater temperature difference than from the temperature of the storage unit of a solar heating system. The heat pump thus uses more electric power without a solar heating system than with it.

The favorable possibility of combining a solar heating system with a windmill to help tide over the long cold and cloudy, but probably windy, periods, was discussed in chapter 3.

The most widely used solar energy system is a modest one that provides household hot water. One form consists of a solar panel on the south-sloping part of the roof connected with coils inside the hot-water tank in the house. The circulation can be provided by a small electrically driven pump, permitting the hot-water tank to be in its normal place in the basement. The hot-water tank could also contain the usual electric resistance heater with its thermostat set to function only when the solar panel fails to keep the water warm enough.

It is possible to avoid the necessity of a circulating pump by placing the solar panel near the eave of the roof and the hot-water tank at a higher elevation just under the peak of the roof. The tendency of the cooler water to descend and the warmer to rise then provides a gravity drive for the circulation. This requires longer plumbing units to conduct the tap water up through the hot water tank.

Simple solar hot-water systems are popular in some parts of the world, especially in Japan and Indonesia, where plumbing is more limited than in the United States. There were tens of thousands of them in the southern states in the 1920s, before natural gas became cheap and convenient enough to be preferred for this purpose. Reintroducing them in much greater numbers now would be one of the simplest ways we could reduce our need for electric power.

Air conditioning has become such a widely accepted necessity for summer comfort that one cannot think of dispensing with it in the interests of fuel economy. One must provide for it and solar energy is capable of doing so. The cooling unit used in the usual air conditioners and refrigerators contains a heat pump that pumps heat from a lower inside temperature to a higher outside temperature. This is just the opposite of a heat pump for home heating, where heat is pumped from a cooler outdoors to a warmer indoors. The most cost-effective way to use a heat-pump system is to make it perform both functions and be useful year round, for heating in winter and cooling in summer. In this case further savings are possible if it serves merely as an auxiliary to a solar-panel system in winter.

Space cooling can also be provided by the direct use of solar energy. It may at first seem enigmatic that one can cool a house by applying the sun's heat, but one should recall that gas refrigerators, employing a gas flame as a source of heat, have been on the market for many years, though they are not popular where electric power is available.

Refrigeration depends on two important facts of nature. The first is that heat is given off when a gas condenses into a liquid, and, vice versa, heat is absorbed when the liquid vaporizes. One may visualize this by thinking of heat as the energy of molecular motion. In a gas the molecules are free to dash around and they give off some of this energy of agitation when they are confined by the attracting forces in a liquid. Heat has to be supplied from the outside to free them and set them in agitation again when evaporation or boiling occurs, as when a kettle is put on the stove to boil water. The second important fact is that the vapor pressure of a liquid, that is, the pressure that vapor evaporating from the surface can exert, depends on the temperature: it is higher for higher temperatures—how much higher depends on the liquid.

An ordinary electric refrigerator contains one working fluid, or one liquid-vapor combination such as freon, and a pump to provide a difference of pressure. The high-pressure part of the circulating system includes a radiator outside of the refrigerator box where the vapor condenses to liquid and gives off heat, while the low-pressure part has coils inside the box where the liquid boils into vapor and absorbs heat, leaving the coils cold. The high-pressure and low-pressure parts are connected by one leg of the circuit containing the pump and another with an expansion valve restraining the flow. That is essentially how any power-driven "heat pump" works, though some are complicated by having an absorbing fluid propelled by the pump. In summary, the pump is used to create a pressure difference so that a fluid condensed into liquid form at high pressure, but kept from getting too hot by circulating air or water, may be evaporated at the lower vapor pressure of a lower temperature. At this temperature the absorption of heat by the evaporation produces a cooling effect.

In one type of gas refrigerator, the pump is replaced by two tanks of another liquid in which the vapor can dissolve. They are at different temperatures and thus have different vapor pressures, so they create a pressure difference just as a pump does. The pressure difference is also maintained by the weight of the column of solution in one of the tanks. One good fluid choice is ammonia and a concentrated water solution of a salt, such as sodium thiocyanate, which dissolves ammonia. At a fairly high temperature this has a good vapor pressure of ammonia and almost none of water. Figure 33 illustrates the principle of one type of operation. The warm-solution tank is heated either by a flame or, preferably, by solar energy and heat is removed by circulation of air or water near room temperature both from the condenser and from the cool solution. The heat conduction pipe between two heat exchangers helps to reduce some of this heat loss and improve efficiency. The circulation of the solution through the tanks is by gravity-driven thermal convection, since the solution is cooler on its roundabout path outside the warm tank than in it.

The more commonly used refrigeration cycle based on a water solution of lithium bromide is somewhat more complicated but the same in principle, employing the weight of a liquid column rather than a pump, to maintain a pressure difference corresponding to a difference in vapor pressure of a solution at two different temperatures. The pressure difference occurs in a U-tube with partly vaporized solution in one leg, and liquid in the other, condensation taking place at the bottom of the U. Another scheme uses solar power to drive a turbine to power the pump of a common refrigerator directly without intervention of electricity.

These processes cool continuously with continuous application of heat. There is an intermittent refrigerating process that is somewhat simpler to construct but requires more attention to complete a cycle. First heat is applied to produce a vapor pressure, then the subsequent condensation produces a cooling effect. This

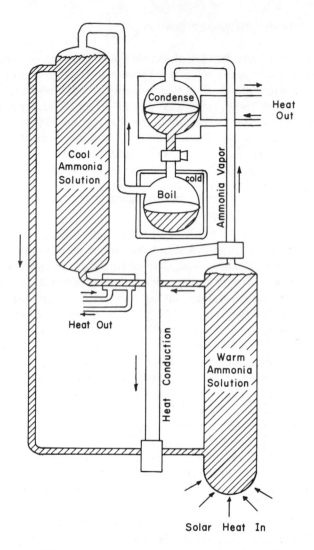

Fig. 33. The principle of refrigeration by the application of solar heat. Ammonia is vaporized from the heated lower tank at the higher pressure of the bottom of the upper tank. It is condensed at that pressure, giving off heat that is carried away. It passes through a pressure-reduction valve and at the lower pressure of the top of the upper tank it boils, absorbing heat from the cold chamber. It then dissolves in a cool solution having a lower vapor pressure.

is promising for making ice in a simple tropical economy, where labor is more abundant than capital, but is not applicable to modern air conditioning needs.

Solar air conditioning is a very attractive use of solar energy because the need is greatest when the supply of solar energy is greatest. It is, indeed, a means of combating excesses of solar heating of the atmosphere, and avoids the tendency of fuel-driven air conditioning to make a hot day hotter. It is useful in daytime without any storage facility, but it may also make use of the solar collection and storage units of a solar heating system.

Besides the use of solar energy to replace the burning of fuels and reduce loads on electric power systems in this country, there are several other uses of solar energy particularly relevant in arid southern countries with less developed economies. Among them are desalination of water by distillation, solar cooking, and dwelling heating and cooling by simpler and more cumbersome methods than are apt to be used extensively in this country. Whereas here we look to saving fuel and electric generating capacity, there, they may save the country's soil itself, by avoiding excessive gathering of wood and grasses for cooking in overpopulated regions and consequent erosion.

A significant example of this sort is related by Professor Farrington Daniels of the University of Wisconsin, a pioneer in promoting uses of solar energy (1971). In a Mexican village where women went to hills twelve miles away to gather firewood for cooking and carried it home on their backs, the people were taught to make four-foot-diameter parabolic mirrors capable of boiling a quart of water in ten minutes. At first these were enthusiastically received but shortly thereafter abandoned, partly because the practice lost prestige when the women learned that women in the United States do not cook that way. Here may be a lesson in facing the problem of providing large power sources for the underdeveloped world. Even if wind power, for example, is uniquely favorable in other countries, we may have to develop and use it first in the United States.

Large Solar Steam-Electric Systems

Solar energy may be harnassed to produce electric power by direct use of sunshine to generate steam to power a steam engine that drives an electric generator. To do this efficiently, higher temperatures are needed than for solar space heating, and this requires focusing of the sun's rays. Focusing is accomplished either by large lenses or by mirrors, by refraction or by reflection. Reflectors are the more practical choice for covering large areas, shiny metal, or metalized surfaces. Two principal types of systems have been proposed: a system using long parabolic mirrors that are approximately cylindrical in shape; and a solar tower system that employs a large number of flat or nearly flat mirrors.

The cylindrical mirror system, and its equivalent using spherical mirrors, have been demonstrated several times. Probably the earliest was at the Paris Exposition in 1876 and the largest was at the Shuman-Boys plant in Cairo, Egypt, in 1913. The latter may be seen as an early prototype of a large-scale system being considered today. The steam was produced directly in pipes at the foci of cylindrical mirrors (see fig. 34) and drove a fifty horsepower steam engine used to pump irrigation water from the Nile River. It was built by American initiative.

The main advantage of cylindrical focusing is the simplicity of tracking the sun with it. The cylindrical axis is placed in a horizontal east-west orientation in conformity with the east-west daily passage of the sun across the sky. In March and September, at the spring and fall equinox when the direction to the sun is at right angles to the earth's axis, the sun crosses the sky in a straight line, and a stationary cylindrical reflector stays in focus all day long. To appreciate this, think of the sun's rays as made up of a component parallel to the cylinder axis and another normal to it. It is only the normal component that is involved in the focusing. At other times of the year the sun's passage is a slightly curved line—a section of an almost-flat cone—and a slight daily rocking is required as well as a gradual adjustment during the year. In the 1913 demonstration the cylindrical mirrors rolled on actual rockers (fig. 34), but in modern versions the mirrors rotate about an axle supported above ground level. The mechanism for this rocking in one dimension is much simpler than the more accurate heliostats required to track the sun in two dimensions. It is simpler also because the same orientation adjustment applies to all the cylindrical reflectors of a large solar farm.

Ideas for improving the efficiency of the heat-absorbing pipe along the focus are being proposed and developed mainly at the University of Arizona by the Meinels (1972), and at Minneapolis by a university-industry team (Eckert, 1973). The main improvements over a simple pipe acting as a boiler are, first, more efficient absorption of the incidental radiation and, second, a heat pipe along the axis inside the absorbing surface to conduct the heat to the end of a long unit, where it is employed in a compact boiler. The heat pipe involves evaporation and condensation of a circulating fluid. Its outer surface is coated with special materials developed for efficient absorption and low radiation. In one proposal this is surrounded by a plastic tube that is internally coated with a reflecting surface over its top half or more, but kept transparent on the side facing the cylindrical reflector, so that much of the light that is not absorbed when first hitting the absorber is reflected

Fig. 34. One of the collectors of the solar steam-power plant built in Meadi, Egypt, in 1912–13 by Shuman and Boys of Philadelphia. It developed 55 horse power for irrigation pumping.

back to have another chance, and radiation loss is kept low. Partial evacuation of the outer tube can also reduce conduction losses.

The steam from a large number of the little boilers at the ends of the heat pipes can then be piped to a central steam engine to produce electric power. In this way, square kilometers of arid country where the sun is bright could be covered with solar energy collectors to feed power into the main electric lines. Because of access corridors and in order to avoid reflectors shadowing one another in winter when the sun's path is low, the reflectors will cover only one-third to one-half of the area occupied, depending on the latitude. This leaves room for growth of whatever vegetation may be possible in the arid country, and perhaps enhances growth by providing shadow without reducing moisture collection. The units may be of sufficient size and height to allow grazing among them.

The solar tower concept is illustrated in figure 35. In it there is a heat receiver at the top of a tall tower, with a turbine and electric generator either in the tower or on the ground beneath it. The heat receiver is either a steam boiler or a hot-air chamber. Sunlight is reflected onto the heat receiver by a huge array of flat or nearly flat mirrors on the ground around it. This system demands that each of these mirrors must track the sun, not only daily but minute by minute throughout the sun's passage across the sky. The great

Fig. 35. Artist's concept of a power tower. (*Courtesy The Boeing Company.*)

advantage compared with the cylindrical system is that all the power is produced at one place, with the sun's rays rather than pipes bringing the energy there.

This makes the power tower the more attractive concept if the problems associated with it can be solved. A boiler capable of withstanding and efficiently using the concentrated heat has yet to be developed and the cost of the thousands of heliostats of a large system is not yet established. There is a trade-off between the cost of the many heliostats and of a high seismic-resistant tower and the central-boiler problems of one system and the extensive piping from the diffuse steam source of the other. The decision has been made, as already mentioned, to give the lion's share of the federal funding to the power-tower development, leaving only a relatively small effort for the cylindrical alternative.

Unlike the steam-generating solar plants, the hot-air power tower works on the principle of a turboprop aircraft engine. The intake air is pressurized by a small fan and after being expanded by heating the air drives a larger turbine. If this type of power tower can be made efficient enough it will have the advantage of not requiring cooling water.

The power-tower form of solar energy may have in common with nuclear power the fact that it is descended from military technology, for it is very similar to a military invention of Archimedes during the Second Punic War more than two thousand years ago! When the Roman fleet came into the harbor of Syracuse, so the legend goes, a thousand soldiers lined up along the quay focused the reflection of the sun from their burnished metal shields onto the rigging of one ship after another, setting them afire.

As with wind power, the drawback here is that solar energy is diffuse and very large areas must be covered to provide as much power as is generated more compactly by fuel-consuming power stations. Figuring direct bright sunlight at 1 kilowatt per square meter, the fraction of day in sunlight at one-fourth, and area coverage and steam power efficiency each at one-third, we can estimate the electric power yield as 30 megawatts per square kilometer. Thus about a 6-kilometer (or almost 4-mile) square would be required to provide the 1,000 megawatts that would correspond to the output of a large fuel-consuming power station if it were to operate continuously at full capacity (or a 2.8 mile square that would correspond to such a large nuclear plant at a more realistic 55 percent plant factor). Obviously this represents a very large area of reflecting surface and the machinery to go with it.

The expense of the system, both in terms of money and materials, is determined partly by the necessity to build the heat-collecting structures strong enough to withstand the strongest winds. This places it at considerable disadvantage compared with wind power because a wind turbine of comparable power output is not nearly so large in area and is designed primarily to deal with wind, whereas this is an extra requirement of a solar-thermal plant.

One important question about the practicability of such large-scale solar power is the extent to which the reflecting surfaces may be made to retain their reflectivity in the blowing dust of a desert environment without a need for too much cleaning.

The power tower, as it is called, is the favored solar-energy project of the ERDA-DOE solar energy program, being funding with more than 25 percent of the entire solar budget (compared with about 8 percent for wind power) (Hildebrandt, 1977; Metz, 1977d). Four aerospace companies each hold a contract for developing a different type of heliostat-mounted mirror unit. These units are quite large having about thirty-seven square meters of reflecting surface, equivalent to a square six meters or twenty feet on a side (Metz, 1977). One of the four types of heliostat is quite light, employing a reflective-coated thin film protected from dust and the weather by an inflated plastic bubble (fig. 36). The success of the program may depend on this as an economical way to withstand wind storms.

The decision to place greater emphasis on the power tower than on any other solar-related technology, such as wind power, was probably based on the expectation that a power-tower electric generating plant will be most similar to a nuclear or coal-fired central station to which the authorities are accustomed, a large generating plant with one large boiler and turbine (Metz, 1977d). As was the case with nuclear power, the development and demonstration is to proceed in a succession of plants of increasing size. The first is a 5-megawatt-thermal (about 2-megawatt-electric) test facility due to be completed in 1978 and the second is to be started in that year, a 10-megawatt-electric pilot plant at Barstow, California. These are to be followed by a 100-megawatt electric plant in the mid-1980s and a prototype commercial plant of about that size or larger in the next decade. The 10-megawatt plant is expected to cost about $12,000 per installed kilowatt, including some develop-

Fig. 36. One form of sun-tracking mirror for the power tower, protected by a plastic bubble. (*Courtesy The Boeing Company.*)

ment costs. The boiler is as much in need of research and development as is the mirror-heliostat unit and if these developments
succeed in reducing the costs sufficiently, this schedule of development and demonstration would lead to extensive use of this power
source (the possibility of 40,000 megawatts has been suggested) by
the turn of the century.

The 100-megawatt plant is to cover about a square mile and
will require a tower about a thousand feet high. It is not considered
economic to build wind dynamos nearly that high even though
there are very favorable and relatively steady winds at that height.
If this seems inconsistent or suggests a bias, it must be appreciated
that there would be only about one-tenth as much power per tower
with wind power and the tower with its wind turbines would present a larger area to storm winds.

The land usage of a solar-thermal plant is much different from
that of a wind-power system, being more concentrated and perhaps permitting grazing, but precluding agriculture. A wind power
system is spread out over a much greater area but so sparsely as
not to interfere appreciably with agriculture. If we consider putting
2-megawatt windmills a half mile apart on the four-mile square that
might provide the equivalent of a continuous 1,000 megawatts of
solar-thermal power, there would be about sixty windmills, which,

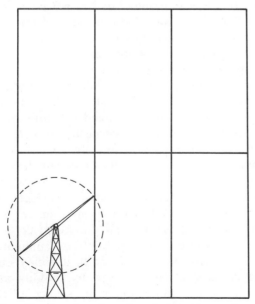

Fig. 37. Comparative size of a wind turbine and a part of a solar farm producing the same average power.

with a capacity factor of 35 percent, would provide the equivalent of a continuous 40 megawatts. This is only 4 percent of the solar yield of that same area. Thus, twenty-five times as much area would be sparsely covered by an equivalent wind-power system. In this estimate, one 2-megawatt wind turbine is equivalent to the part of a solar system in a square 170 yards on a side, or about the area of six football fields. This statement is illustrated in figure 37, to give an idea of the area occupied by an array of solar collectors as compared with the size of a 2-megawatt wind turbine. The blade span is 200 feet, a little more than the 160-foot width of a football field. The area of the reflecting surfaces alone, if packed close together, is enough to cover only two of the football fields. Regarded thus, it might seem that covering two football fields with reflectors would be an easy alternative to constructing a wind turbine so large that, when laid on its side, it would appear to occupy about one football field. However, it must be appreciated that those reflectors must be mounted to be rotatable and still strong enough to withstand high winds, whereas the wind turbine is a spidery structure with a small area exposed to the wind in comparison with the space it appears to occupy.

Solar-Cell Generation of Electricity

The electronics of solid-state semiconductors has advanced so rapidly that the function that would have been performed thirty years ago by a hundred vacuum tubes are now performed by a tiny chip smaller than one's little fingernail. Moreover, the technology of making these tiny chips and their associated components is so advanced that one may buy a miraculous pocket calculator for ten dollars. A similar development of a solid-state product and of the methods for producing it economically are needed before solar-cell generation of electricity may achieve industrial importance. When and if this happens, it can contribute substantially to the large-scale needs for commercial electric power as well as on a smaller scale to home electric needs.

The functioning of some crystalline substances as electric semiconductors depends on the fact that electrons may move in them with only certain specific energies, in what are called certain energy bands in the energy spectrum. It is possible for sunlight, as it is absorbed, to lift an electron from a lower to a higher energy. The electron can then expend this energy by moving around through an external circuit and can re-enter the semiconductor in the lower energy band, ready to be lifted again. The energy so expended in the external circuit provides electric power, directly from the absorption of sunlight. Of course very many electrons are involved in this process at once, pushing each other around the circuit as an electric current.

The technical practicability of this method of generating electric power is attested by its success on space missions that are entirely dependent on solar cells for their power. Very substantial amounts of power are needed for life support systems in manned missions, for guidance systems, and various experiments, as well as for communications with Earth on all space journeys. The large panels of silicon solar cells unfold after the space vehicle has left the earth's atmosphere and are oriented to face the bright sunlight of space in our part of the solar system, free of the atmospheric absorption we have on earth. The technical practicability so demonstrated, however, says nothing about economic practicability. The present very high cost of these silicon solar cells is a small part of the cost of a space mission. Only a few applications in very special circumstances have been found on earth.

Cadmium sulfide photovoltaic cells are somewhat less efficient than silicon cells but are already considerably less expensive, around $20,000 per kilowatt as compared with well over ten times that for the cells that were used on the early space missions. They function even when made extremely thin, thinner than a human hair. Some progress has been made in developing a process of producing the material in the form of thin ribbons, encouraging hope for the future that this may be done inexpensively in wide thin sheets that may still be mounted in a rugged way. They would then be deployed over large areas of arid land, in the manner described for cylindrical reflectors. They would have a great advantage over reflectors because there is no need to generate steam and therefore no need for cooling water.

The electric power is generated directly as direct current and in most cases would be transformed to alternating current by electrical means for long-distance transmission. Even though there is no energy loss by cooling water, efficiencies of photovoltaic cells are still somewhat less than for steam-electric systems. It is hoped that they may be improved.

One advantage of the prospect of covering large areas with photovoltaic surfaces that do not require focusing of the sunlight is that they also function in diffuse light from a bright sky with a light overcast. Another advantage is that no machinery for tracking the sun is needed.

An experimental solar house at the University of Delaware (Boer, 1973) depends on solar panels that provide both hot air and electric power. They are again a sandwich construction with a transparent plastic sheet on top, an absorbent layer in the middle and air ducts between that and an insulation layer on the bottom. In this case the absorbing layer consists of cadmium sulfide solar cells providing the electric power that charges storage batteries while the hot air warms a thermal storage unit, and a heat pump boosts the heat to a higher temperature for hot water and space heating. It is estimated that with quantity production of the critical units such a house would cost only 10 percent more than a conventionally heated and powered house.

Ocean-Thermal Generation of Electricity

The strong heating effect of nearly vertical sunlight on the ocean in the tropics, compared with the weaker effect near the poles where sunlight is slanted, creates temperature differences in the ocean that can be an important source of industrial power. Once again the source is diffuse and large structures will be needed to harness it. The cold water of the more polar regions flows in under the warm water nearer the surface in the tropics. Since the warm and cold sources are not very far apart it is practical to make a heat engine work between them. This system takes advantage of the surface of the ocean as a ready-made solar collector but the sun's heat is dispersed through a rather large volume of water that must be passed through an evaporator to convert the heat to useful energy.

The maximum efficiency of any heat engine is proportional to the temperature difference between the heat source, at temperature T_{hot}, and the cooling medium, at temperature T_{cold}. Specifically, the Carnot expression for the ideal efficiency is $(T_{hot}-T_{cold})/T_{hot}$. Here the temperatures are measured on the absolute scale, starting from absolute zero at $-273°$ C or $-460°$ F. Expressed in degrees Fahrenheit, the formula for the ideal efficiency may be written (Temperature difference)$/(T_{hot} + 460)$. The boiler in a fuel-consuming steam plant is at high pressure so that boiling can take place at a high temperature, T_{hot}, to make a large temperature difference for the sake of high efficiency.

In harnessing ocean-thermal differences, T_{hot} and T_{cold} are determined by the ocean temperatures available. In favorable tropical or subtropical locations T_{hot} is around 80° F or a little more and T_{cold} around 40° F or a little more, the difference thus being about 40° F, at some places as high as 45° F. This means an ideal maximum efficiency, according to the formula, of about 40/540 or 7.4 percent. In practice one expects to achieve no more than about 2 or 3 percent efficiency in the conversion of heat inflow to electric power. Because the heat input available from the solar heating of the ocean surface is so vast and free and the heat sink is nearby, this is nevertheless a practical proposition.

There are two rather different methods for harnessing ocean-thermal differences. The older one, the Claude cycle or open cycle, utilizes the vapor pressure of seawater itself as the working medium and has been demonstrated to be practicable. The other method, a closed cycle known as the Rankine cycle, uses a working fluid with higher vapor pressure at the temperatures available. This cycle is favored for future development in expectation of higher efficiency but does not yet have the advantage of having been put in practice. Compared with the prospects of wind power for large-scale development to help meet our energy needs soon, it is this lack of experience that puts ocean-thermal-difference development at a disadvantage and leads to the emphasis on wind power in this book. In the case of wind power the favored method has been demonstrated as practicable in several countries over the past half century and has failed only the economic test in an era of cheap fuel, whereas the favored method of ocean-thermal energy conversion (OTEC) has not been so demonstrated. It is to be hoped that it soon will be; there are reports that Japan plans to lead the way in actual deployment. Research and development particularly of materials, fouling problems, and design are in progress in this country with funding from DOE.

The idea of ocean-thermal energy conversion with a suitable working fluid was originated by d'Arsenval in 1881. The technical feasibility of the open-cycle system was demonstrated by Claude with an installation on the south coast of Cuba in 1929. It was a remarkable achievement at the time. The electric power generated was 22 kilowatts with an overall efficiency less than 1 percent. The hot and cold water were conducted through long pipes to the machinery ashore. With the limited technology and cheap fuel of that time, there was then little prospect for economic feasibility. A larger installation with two units totaling 7 megawatts was constructed on the Ivory Coast by the French in 1956 but encountered troubles and was abandoned.

For the closed cycle a working fluid with boiling point near and preferably between T_{hot} and T_{cold} is chosen so that the working pressure may be near atmospheric. Ammonia and propane are examples. Large heat exchangers are required, an evaporator to transfer heat from warm seawater to the working fluid and a condenser on the other side of the turbine to transfer the exhaust heat back to cold seawater. An important part of the design problem is to make great quantities of seawater flow through these heat exchangers with a minimum expenditure of energy propelling it. This calls for a very large-caliber pipe long enough to connect the deep level of the cold water with the near-surface level of the warm water. Presumably the main body of the power plant will be on or near the surface and the long pipe will draw cold water up to it. The use of a working fluid requiring the awkward heat exchangers has the advantage that the higher vapor pressure drives a much smaller turbine than is required with water vapor.

The preliminary engineering design shown in figure 38 gives an idea of the type of structure that may be involved, submerged except for the access tower containing crew quarters. The huge heat exchangers here are external to the main hull and are seen to be much larger than the turbogenerators between them. Warm water is pumped in through the screens at the top. The cold-water intake pipe seen at the bottom, surrounded by flotation cylinders, is to be assembled by letting down telescoping sections. It is proposed that in the first test version the pipe should extend down only 1,000 feet, establishing a temperature difference of about 20°F, but in the final version it would be nearly 2,000 feet deep. Some of the research projects sponsored by DOE are focused on a more primitive pilot with smaller heat exchangers inside a circular main hull that floats on the surface, using conservative hull design and construction methods developed in the exploration for North Sea oil. Probably concrete will be widely used in such construction, in keeping with concrete-ship experience.

The most favored sites at least for early deployment of this type of plant are the Caribbean Sea and the Gulf of Mexico, where it can either be anchored or left free-floating with a weak propulsion capability. The latter plan avoids mooring costs and the propulsion can be supplied economically by directing the warm-water and cold-water exhausts appropriately, since the kinetic energy of the exhaust flow must be dissipated somehow anyway. Although the preliminary version in figure 38 with its 1,000-foot depth, is planned for mooring with a power cable to shore, it is contemplated that others would have no direct power conduit to land but that useful transportable materials, either fuel or else ammonia for use in making fertilizer, will be produced to be shipped ashore.

If the plant floated in still water it would be cool the warm water around it, reducing its efficiency. However, the ocean has different currents at various levels. The deep cold-water intake pipe acts as a drag to make a warm current flow past the upper part. Furthermore, the discharged cold water tends to sink back down to whence it came.

Power module

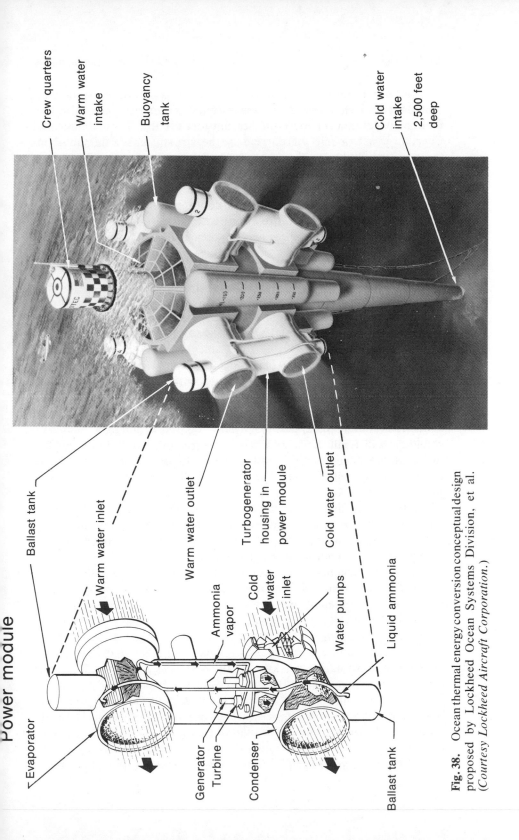

Fig. 38. Ocean thermal energy conversion conceptual design proposed by Lockheed Ocean Systems Division, et al. (*Courtesy Lockheed Aircraft Corporation.*)

Quite a different design using the same principle is being promoted by a University of Massachusetts group for exploiting the thermal difference in the Gulf Stream just off the Florida-Georgia-Carolina coast. While the current of the stream makes demands on the mooring integrity, the current can be used to propel the large volumes of water to and through the heat exchangers eliminating much of the need for pumps. This possibility, and because the stream brings a plentiful heat source to the system, along with proximity to the coast of the Southeast permitting electric power conduction directly to shore, are important advantages of this Gulf Stream system. A serious disadvantage is that the temperature difference (roughly 32° F) available is less than that in the more southerly seas (40° F) and this means lower efficiency. The advantages seem to outweigh this disadvantage but there is another drawback that may be decisive. The speed of the natural flow through the heat exchangers may not be great enough to suppress fouling of the surfaces.

One proposed version of an ocean-thermal plant for the Gulf Stream is shown in figure 39. The large array on the top is a multi-tiered array of heat exchangers, serving as the evaporator, through which the upper layers of the Gulf Stream flow to supply input heat. The remaining apparatus, including the condenser heat exchangers and turbine, is enclosed in twin submersible hulls resembling submarines. From the bow a long and large-diameter inlet tube reaches down to scoop up cold water from the depths (McGowan, 1976).

There is a clear advantage to having the warm water simply flow through the evaporator structure, rather than pumping it, as in figure 38. This appears, however, to leave the huge heat exchanger more susceptible to wave damage in a severe storm. The whole structure is normally submerged except for the slender shafts supporting the access decks, and the hazard is avoided by deeper submersion at the approach of a hurricane.

It is too early to put much credence in estimates for capital costs, but they range from approximately $650 per kilowatt electric for a floating system using ammonia and a temperature difference of 36° F to $2,700 per kw$_e$ for the 190-Mw$_e$ system of figure 38 (or half that cost in quantity production) using propane and 18° F. The Gulf Stream system of figure 39 is nearer the lower figure, employing propane and a temperature difference of 32° F and including wiring to shore. The lower figures correspond to around 15 mils per kilowatt-hour, roughly half of present municipal electric rates.

The open-cycle system that uses seawater vapor directly is

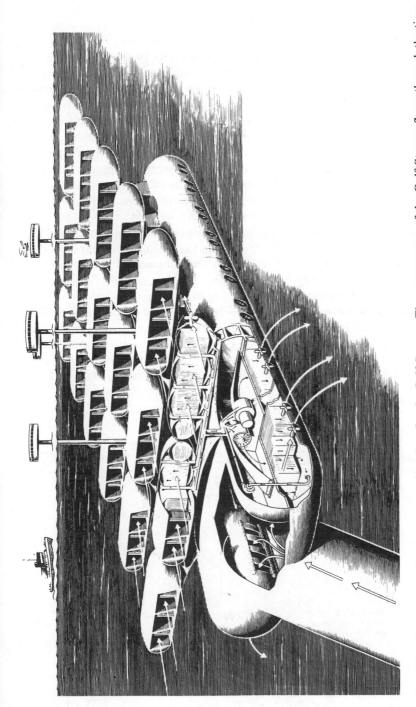

Fig. 39. Ocean thermal energy conversion plant for the Gulf Stream. The warm water of the Gulf Stream flows through the tiers of heat exchangers at the top. (*Courtesy Jon McGowan.*)

shown schematically in figure 40. At first sight there seem to be so many pumps and turbines that one wonders if there is a net gain in energy conversion. The answer depends, as in other power and refrigerating systems, on the fact that much less energy is required to pump a certain amount of fluid against a given pressure difference if the fluid is in its compact liquid form rather than in its expanded vapor form. A much smaller piston, for example, is required to push the compact fluid against a pressure than to push the much larger (even a thousand times larger) volume of vapor. Conversely, the larger volume of vapor passing through a turbine generates more mechanical energy. Thus only a small fraction of the energy generated by the steam turbine in figure 40 is used by the electric motors that help drive the water pumps.

In order that the working fluid shall be water vapor unimpeded by the presence of much air, the steam turbine and the upper parts of the evaporator and condenser must be in a partial vacuum containing essentially only water vapor. This means that the input water must be greatly reduced in pressure and the exhaust water pumped back up to pressure. Fortunately for the energy balance, the pump that is restoring the pressure can be driven largely by a turbine driven by the input water being reduced in pressure. Under ideal circumstances after balancing small differences on the evaporator and condenser side, this drive would be almost sufficient but

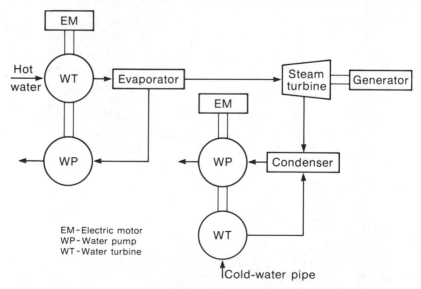

Fig. 40. Schematic diagram of the Claude Cycle, employing the vapor of seawater directly to drive the turbine.

motors are required to make up for inefficiency in the process. Only a small fraction of the warm input water is evaporated, for heat must be extracted from the rest of it to accomplish the evaporation, so large volumes of water must be pumped as efficiently as possible. Compared with the closed cycle, this open cycle involves very simple heat exchangers, essentially just tanks capable of withstanding atmospheric pressure, but demands much more in the way of pumps and steam turbine size.

While proposals have been submitted to go ahead with the construction of a large ocean-thermal-differences demonstration power plant such as that in figure 38, the Department of Energy in 1977 seems to approach the development more cautiously and hence slowly. The plan is to start with several years of tests with heat exchangers at sea before considering authorizing construction of the proposed large unit (Metz, 1977c).

A large old barge known as the Hughes Mining Barge will be used as an ocean test platform (Seamans, 1976). It will carry a six-foot diameter cold-water pipe and four heat exchangers to be tested, 1-megawatt-scale evaporators and condensers, quite sizable but much smaller than those in the large unit that has been proposed. The barge is to be moored, at least initially, within eighty miles of shore. Dr. Seamans says of it, "ERDA's program is intended to find out by 1984 whether a large, commercial-scale floating power plant of this type is economically feasible."

In the case of wind power, the federal program has progressed more slowly than is appropriate in view of the urgent need, its first four years having seen completion of but a single wind dynamo, almost a duplicate of one built abroad a dozen years earlier and much smaller than the one achieved in less than two years in Vermont thirty years ago. The plan in the ocean-thermal program to start with several years' experience on a test platform, rather than proceeding directly with the final engineering and construction of a practical power plant, represents a similarly cautious approach.

This may seem too cautious in view of the need for rapid progress, but such a careful approach is more reasonable for ocean-thermal projects than for wind power because the test platform work will be exploring new territory, not just repeating what has gone before. The results obtained concerning biofouling in heat exchangers will be more specific for the purpose than past experience with biofouling of ships' hulls and will permit greater confidence in designing a practical power plant at a later date than could be done now. Yet, as in the case of wind power, proceeding forthwith to build a full-scale power plant, such as in figure 38, with some test platform work, would lead more directly to the goal of early achievement of economic electric power. Even if the Mark I power plant should be far from economic and run into technical trouble, it would give experience in construction and operation and the information about troubles would be more specific and helpful in designing Mark II than results from scaled-down tests could be.

Thus both large-scale wind power, the solar-related source that is most nearly ready to proceed, and ocean-thermal power which is next in line, could make more rapid progress toward important power production if they were given higher priority in national planning.

Wave Energy

Ocean waves contain a lot of energy that is derived from the winds, so the ocean surface can be viewed as an enormous collector of wind energy as well as of direct solar energy. Wave motion consists of both vertical and horizontal motion. Not only does the water near the surface move up and down with the passage of the crest and trough of a wave, but in a gentle wave without whitecaps it also moves forward at the crest and backward at the trough.

The individual particles undergo an approximately circular motion, moving up as the crest approaches, then forward at the crest, down as it recedes and backward in the trough.

It has long been proposed that the vertical motion could be converted to useful energy by taking advantage of the differential motion between a float and an anchor. In a recent suggestion, S. H. Salter (1974) proposes to derive electric power from the horizontal motion. If two sets of vertical panels acting as paddles are spaced so that the paddles of one set are at the crests of the waves while those of the other set are at the troughs, the to-and-fro motion of the two sets can be converted to mechanical energy and thence to electricity. He claims that a unit the size of a supertanker suitably located could generate a yearly average of 50 megawatts. This is about a quarter of the output of a proposed ocean-thermal energy conversion plant that has about the weight of a supertanker, so according to these rough estimates it appears that exploitation of wave energy is apt to be less cost effective than that from ocean-thermal differences. There is nevertheless active interest in wave power in Great Britain (Hawkes, 1977).

Organic Waste Conversion and Biomass Farming

The life on this planet derives energy from the sun by photosynthesis. The process is direct for plants and indirect for the animals that feed on plants. The growth of living things depends on building complex substances such as macromolecules and cells out of simpler ones. After death these eventually decay, releasing some of the energy stored in them. In many cases the decay is aided by microorganisms that feed on the released energy. The potential of releasing energy in what is left may be exploited to help meet our energy needs. The marsh gas that rises from the decaying vegetation beneath a pond is methane, a very useful fuel and the principal ingredient of natural gas that was made by a similar process eons ago.

A lot of organic material for energy conversion is available as waste from our present growth patterns and way of life, without being specially grown for the purpose.

There is a lot of rotting of wood on the forest floor that is too diffuse to be exploited economically. The sources of organic waste that are probably most useful are those that are already collected as byproducts of our way of life. These include city trash, sewage, manure, and wood waste at sawmills. Also, specific crops could be grown for the purpose of conversion to fluid fuels. The prolific water hyacinth growing on ponds fed with nutrients from organic wastes and separated from its water by centrifugal methods is a promising possibility. It is thought that these and land crops such as grasses and trees could be converted to liquid fuels to produce energy from incident sunlight at overall efficiencies of less than 1 percent to perhaps 3 percent. This gives much less energy per unit of area than does steam production using mirrors, but does represent a less expensive way to cover area with solar collectors.

There are three principal methods of decomposing wastes to obtain useful energy, bioconversion, pyrolysis, and direct burning. The decay of vegetable matter at the bottom of a pond and the treatment of sewage in a municipal plant both depend on the bioconversion process known as anaerobic digestion. This is decay promoted by certain bacteria in the absence of oxygen. In some sewage plants the methane produced is wasted, in others it is used to heat the incoming sewage to a temperature that promotes rapid disintegration. This is already a reasonable use of solar energy for a specific purpose. However, the usual sewage plant conditions are arranged to maximize breakup of the sewage. As an energy source they would be arranged to maximize the output of methane for other purposes and still break up the sewage.

Bioconversion of farm wastes, and particularly the huge available quantities of feedlot manure, would involve the construction of large vats closed at the top for collection of the gas produced.

Pyrolysis is a process of decomposing organic matter at high pressure and high temperatures in the absence of oxygen. It is similar to the way coke is made from coal and indeed a burnable solid somewhat similar to coke, known as char, is one of the products. The more important products are oil and gas. The gas may be used to provide the heat for the process itself, leaving oil as the main useful product. There are three types of pyrolysis under development or in a pilot-plant stage, two of them (by the Bureau of Mines and Garrett) making oil as the main product, and another (by Monsanto) producing mainly a relatively low grade of fuel gas for use in space heating. Both products are quite low in sulfur content as is required of good fuels.

Direct burning of city trash can generate electricity with only modest changes in power plants designed for burning coal and in fact this is done economically in some small municipalities already. The cost includes a cost of separating some parts of the trash that do not go into the furnace but still is considerably less than the cost associated with disposal of the trash as landfill. One drawback is that about 70 percent of the burnable trash is paper and in an ecologically sound economy it would be more desirable to recycle the paper than to burn it for power.

While the economics of bioconversion and pyrolysis remains quite uncertain at this early stage of development, there is one attractive feature that probably will make both methods cost effective when organized on a large scale near the sources of wastes. This is the fact that the wastes must be disposed of anyway, and these two processes greatly reduce the volume requiring disposal. Disposal by landfill is surprisingly expensive, some five dollars per ton, and the cost of these processes may be not much more. Feedlot manure is an ecological menace, because the drainage fouls rivers and changes the life forms in them. To have this valuable biological material properly processed rather than being left to wash away would be a boon not only because it would avoid pollution, but because the bioconversion for the sake of energy would also make good fertilizer in a convenient form as a byproduct. It has been estimated that if all the organic solid waste produced by our economy could be collected and used in pyrolysis to produce oil it could account for as much as 25 percent of our 1970 petroleum consumption or, if used in bioconversion to generate energy, it would supply about 13 percent of the United States energy budget. The part, however, that could reasonably be collected would correspond to only about 2 percent of our energy use (Ruedisili, 1975). This represents an opportunity that should be exploited, but a limited one.

Organic materials for energy conversion can be grown specially for the purpose, but this of course involves great effort in both growing and collecting the materials. Ordinary trees or special fast-growing trees are possibilities. Others are fast-growing plants like the water hyacinth, that grow floating on the surface of still water, or sugarcane, the fastest growing, easily manageable crop on land. The traditional woodlot that grows fuel for fireplaces and wood stoves makes its modest contribution to alleviating our national energy problem even now, but the area of forest that would have to be harvested, or the water and fertilizer requirement for the more concentrated growing of sugarcane, to meet modern energy requirements would be prodigous indeed. Northern-type forests to satisfy this need would have to cover the entire area of the United States, including Alaska. Enough sugarcane would cover about the area of Texas, and, even more limiting, would require more than fifty times as much water as flows in the Colorado River, whose water is already much in demand for other purposes. The annual fertilizer requirement would include 7 million tons of potash, for example (Calef, 1976).

While wood burning by individual households suitably located should be encouraged, it is unlikely that large-scale energy conversion of farmed biomass will prove worthy of the effort and expense required to account for as much as 1 percent of our energy needs.

There is, however, hope that more energy may be obtained economically by burning biomass harvested from the open ocean. A very large aquatic plant, kelp, has been shown to be economically combustible. Kelp is quite plentiful in the Pacific off the coast of California and it is proposed that it could be grown much more plentifully if nutrients were pumped up from deep water. Floating windmills would be ideal to do the pumping.

6

Geophysical Energy Sources

Besides all the wonderful bounty that makes up the life-support system on the beautiful surface of spaceship Earth, basking in sunlight, there is yet another bounty beneath the cool solid surface. The inside of the earth is so hot that much of it is liquid. The solid surface layer is only about thirty kilometers (about twenty miles) thick on the average, a thin crust extending down a mere half percent of the distance to the center of the earth. Beneath it is a thick layer known as the mantle, made of a dark rock called basalt. In this layer the basalt is hot and under high pressure and flows very slowly like a plastic, not like a real fluid. In this condition the basalt is called magma. When the pressure is relieved near the surface of the earth it can flow faster in the form of volcanic lava. The solid outer crust consists both of basalt and of the lighter rocks that form the continents. Most of the earth's surface is ocean and under it the basalt forms the ocean floor. The continents are higher than the ocean floor because, being lighter, they float on top of the slightly plastic basalt. In some places the magma comes closer to the surface than at others and can cause surface manifestations like hot springs. The mantle extends almost halfway down to the center and comprises most of the volume of the earth. Inside this thick spherical shell is the liquid core that may be made of molten metals, mostly iron and nickel, comprising more than half the diameter of the earth. At the very center is the solid inner core, again probably mostly iron and nickel, a little more than a quarter as big as the earth's diameter.

Radioactivity in the inner core is an important heat source and radioactivity in the crust also contributes. The earth may have been hot from its beginning as bombarding dust particles from interplanetary space assembled to form the earth, but the radioactivity keeps it hot as heat flows out through the surface. Fortunately the crust is a fairly good insulator so the heat flows out slowly and does not disturb us. Mining and drilling have shown that the temperature of the crustal rock increases with depth at an average rate of about 17° C per kilometer (or 50° F per mile). This rate is rapid enough that miners in deep mines work in sweltering heat in spite of forced air circulation.

Beyond that, what is known of the earth's interior heat is from indirect evidence. The existence of the plastic and liquid layers is manifested by reflection and refraction of seismic waves. The earth's magnetic field is thought to originate from dynamo action of the convective currents in the liquid core that are driven by heat from the radioactivity in the inner core (Bullard, 1949; Inglis, 1966). Variations in the magnetic field are indicative of motions at great depth. From the liquid core upward the heat is carried partly by conduction and partly by very slow convective currents in the plastic mantle. The continents in the crust have drifted slowly around on the surface of the earth presumably because the plates of the lithosphere in which they are embedded ride on those convective currents. Near the mid-ocean ridges those currents rise close to the seabottom and bring up magma to fill in between the plates as they separate (Burke, 1976).

Geothermal Energy

The practical implication of all this is that there is so much heat in the earth's interior that in exploiting geothermal energy we cannot exhaust the supply. We may temporarily exhaust local concentrations, but they will probably be replenished by heat flow from beneath.

Some local concentrations manifest themselves in spectacular displays such as the geysers and hot springs of Yellowstone and less spectacularly in the steam venting at the Geysers in California. Until now geothermal energy has been exploited only in the relatively easy places where there are surface manifestations. It remains speculative how much can be done beyond those efforts. The situation is similar to that over a century ago when petroleum began to be used rather extensively as an energy source but was exploited only where there was surface seepage of oil. Assessment of reserves at that time would have been rather small compared with what has since been economically exploited by deep drilling and modern methods of exploration. There is a good possibility that exploitation of geothermal energy may follow a similar course. For this reason it should be considered a potentially important resource worthy of more vigorous research and development than it is receiving.

The only commercial geothermal power plant in the United States is the one at the Geysers near San Francisco, where steam from wells managed by Union Oil Company is used to generate electricity by Pacific Gas and Electric. The power generated has gradually increased since 1956 with addition of new wells and is now about 500 megawatts. It is thought that when fully developed the Geysers field could deliver about twice that, or as much as one of the largest fuel-consuming power plants, for more than thirty years and perhaps as long as a century or even longer (Rex, 1971), though eventually the local concentration will be exhausted for practical purposes. It is expected that the entire Imperial Valley in California could yield much more, perhaps 20,000 megawatts.

Individual drilled wells at the Geysers deliver steam for up to 7 megawatts and individual turbines, each fed by several wells, generate about 60 megawatts. The installations with one or two turbines are a mile or two apart. The dry steam, at wellhead is at almost 400° F, much less than the 700° F of a nuclear plant or almost 1,000° F of a modern fossil-fuel plant, so the efficiency is less, in the neighborhood of 20 percent. However, with no cost for fuel, this still generates electricity at a favorable price, about half a cent per kilowatt-hour. There is a cooling problem as in other steam plants, but it is slightly different since the condensate does not have to be returned to produce more steam, therefore it can be used as part of the water that is evaporated in a cooling tower to produce a cooling effect.

Impurities in the steam cause complications that may become more serious in future exploitation of less favorable sites. The high-pressure steam brings up solid particles that are separated out in centrifugal separators so as not to let them damage the turbine blades. Chemical impurities escape as unpleasant gases into the atmosphere and also can corrode piping or clog it with mineral deposits. In fact, an early attempt to generate power at the Geysers in the 1920s was defeated by corrosion before suitable materials were developed.

The pure steam at the Geysers is known as dry steam. There are other places where wet steam is available from within the earth. Where this is used for power an extra process of separation of steam from water is involved, though the possibility is under consideration of ejecting the mixture directly into the turbine. It is also an ecological disadvantage that rather large amounts of highly mineralized water must be disposed of, preferably by reinjecting it into the earth whence it came, to avoid depletion and land subsidence as well as surface-water contamination. With dry steam at the Geysers there is only the relatively small amount of condensate water to be disposed of. Of this, 80 percent is evaporated from the cooling towers and the remainder, carrying what little impurities there are, but containing too much mineral to be released without harm in local streams, is reinjected back into the earth.

It is also possible to generate power at hot springs where only hot water comes to the surface of the earth. These occur where there are large enough cracks in the surface layer that cool surface water may flow down one crack to the hot, presumably porous, layer below and then flow up as hot water through another crack. The flow is maintained by the greater weight of the denser cold water, just like the gravity circulation in the old hot-water heating of a house with a basement furnace.

Exploiting the temperature difference between this hot water and the cool water of a lake or stream or cooling tower involves very much the same technique as was described in chapter 5 for ocean-thermal energy conversion. The closed cycle, using a suitable working fluid between heat exchangers, is preferred over the open cycle and it is reported that a closed-cycle plant is in operation on the Kamchatka Peninsula in the Soviet Union. There is the great advantage that considerably larger temperature differences are available in the geothermal than in the ocean-thermal case, with correspondingly higher efficiency and smaller turbines, but the heat source is more limited.

Such thermal sites are located where the earth's crust is thin and the hot magma below comes quite near the surface. This is apt to happen also near the edges of volcanoes. There are doubtless many places that remain to be discovered where the magma likewise comes close to the surface but gives no surface evidence, there being no cracks to conduct water down naturally. If these areas have porous rock beneath the hard surface, perhaps they could be developed as easily and economically as at the Geysers. If the lower hot rock is hard, it has been suggested that it could be fractured by injecting high-pressure water at something like seven thousand pounds per square inch, followed by sand to keep the cracks open (as is done already in an oil extraction technique) to make way for water either to become steam tapped by another drilled well or to circulate as hot water in an artificial hot spring. It is expected that the cooling of the rock locally would cause it to shrink and extend the cracks further, thus tapping the heat in an extended area. When cooled the water should be reinjected into the earth both to dispose of its heavy mineral content and to reduce the likelihood of land subsidence. Extensive use of this system may raise the question of triggering earthquakes.

This technique could be used almost anywhere, not just at the favorable spots, but at higher cost for drilling deeper or using a smaller temperature differential generating power at lower efficiency. With the average temperature gradient of 50° F per mile, an attainable well three miles deep gives a difference of 150° F, which is potentially useful. If future development makes this economic in comparison with future fuel costs and the capital expenditures of solar energy, the amount of energy that could be extracted is almost limitless, entirely ample for future needs. The answer to the economic question is, however, very doubtful, requiring a long extrapolation from present experience that has exploited only the favorable sites.

Present experience is, indeed, rather limited. The Geysers field has the greatest yield, about 500 megawatts, and there are installations at similar favorable dry-steam sites with outputs almost that large in Italy, where steam vents have been known since ancient times and electric power utilization began early in this century. There are smaller outputs in New Zealand, Japan, and Iceland (Wilson, 1974). Wet-steam is used in New Zealand and there are projects with injected water near volcanoes in Mexico and, as already mentioned, some experience in the Soviet Union (Kuwada, 1973).

In addition to generating electric power, there is also some use of geothermal waters in various countries for home heating and cooling and for industrial processes. As fuels become more expensive, space heating might turn out to be an important use of the earth's heat, for the low temperatures needed could be obtained almost anywhere by drilling to smaller depths than needed for electric plants. This source might team up well with solar heating to reduce dependence on storage.

Until the late 1960s it was thought that there was little more geothermal potential in the country than that represented by the Geysers field near San Francisco, and perhaps that is why there has not been more widespread development. Recent estimates of the availability of as yet undiscovered favorable sites that could be developed at competitive cost using present techniques are more optimistic. The estimates cover a wide range. For the total energy available before the sources might be exhausted, they range from about 5×10^5 megawatt-years (Muffler, 1972), equivalent to total United States electric power (in 1976) for a year, to two hundred times that (Rex, 1972). Other estimates for the generating power that could be installed are from 10^5 to 10^7 megawatts, from 20 percent to twenty times present generating capacity (Wilson, 1974). In any case the potential is great enough to warrant intensive development of methods of exploration to resolve this uncertainty and lead the way to widespread exploitation.

Energy from the Tides

The tides are caused by the motions of the earth and moon as members of the solar system, by the spinning of the earth with the gravita tional pull of the moon and the sun. Tidal energy is thus derived mainly from the energy of the earth's rotation and in this sense is geophysical energy.

The tides along most open coastlines run about a meter high but in some narrow bays or straits, where the tides are amplified by a funneling action, they rise as high as ten meters or thirty-odd feet daily. In these relatively few favorable places they may be harnessed economically. There are actually only two tidal energy electric generating plants in the world, one in the Rance River estuary at the head of Saint Malo Bay on the English Channel coast of Brittany in France and the other reportedly at Kislaya-Guba in the Soviet Union. The one at La Rance generates 240 megawatts out of an estimated 1,500 megawatts potential power output for that estuary. The corresponding estimated potential for another bay on that French coast, at Mont Saint-Michel, is 10,000 megawatts and in Nova Scotia, Canada, the estimated potential of the Bay of Fundy is about 30,000 megawatts, two-thirds of it in the eastern branch, Cobequid Bay, where the tides run through the relatively narrow Minas channel. If the Bay of Fundy tides were to be harnessed to the extent of only one-sixth of their estimated potential, as is done at La Rance, this would be equivalent to replacing about five of the largest fuel-consuming power plants, or, if it could all be harnessed, about thirty of them. This being by far the most favorable site available in North America (and, indeed, the most favorable in the world) it is seen that if developed tidal power could make a real but rather limited contribution to meeting our electric energy needs. The most favorable site for the United States is Passamaquoddy Bay on the Maine–New Brunswick border, where the potential is about the same as at La Rance.

The tide is harnessed by erecting a dam across a bay near its mouth and installing turbines through which the water may flow both ways, as the level in the basin is alternately raised and lowered twice a day. This of course changes the local tidewater ecology, fishing, and navigation rather markedly as the amplitude of the tides is changed. The change may include some favorable effects as the troublesome high tide at the head of the bay is moderated, but it will reduce the extent of tidal flats where the ecology has adapted to growing organisms that feed the sealife. As in the case of conventional hydroelectric power, silting may limit the useful life of the installation to a century or less.

The cost of tidal power may be judged by the experience at the Rance River plant (Gray, 1972). It was completed in 1966 at a cost of about $100 million. While its installed generating capacity was 240 megawatts, its net yearly output of about 540 million kilowatt hours corresponds to a performance factor of 26 percent. This means a capital cost of $1,600 per average generated kilowatt, or, figured at a yearly capital charge of 12.5 percent, two cents per kilowatt-hour, about three times the cost of fuel-generated electric power back in 1966 when fuel was cheap. Thus the economics of tidal power, even with the recent hike in fuel prices, appears to be marginal but it has the great advantage of not depending on non-renewable resources, aside from the long-range silting problem.

7

Nuclear Power from Fission

Nuclear energy produced by nuclear fission has two interrelated aspects. The slow, controlled release of nuclear energy in the reactors of power plants is a very useful source of heat that is converted to electric power. The sudden uncontrolled release of nuclear energy in nuclear weapons is a terrible threat that is precariously suspended over mankind like the sword of Damocles, so terrible that most people chose to dismiss it from their everyday thinking.

The technologies of these two aspects and the political initiatives and the concerns that have shaped their development are discussed at some length in a book by the author, *Nuclear Energy: Its Physics and Its Social Challenge* (Inglis, 1973). If the reader feels the need of background for the present discussion, it can be found there and elsewhere. The social challenge implied is the challenge to mankind to reap whatever benefit he can from nuclear energy insofar as this can be done without undue harm or risk of destruction by its awesome threat.

The main purpose of this volume is to compare the attractiveness of various sources of power to supply national and world needs for large-scale electric and other power. Such a comparison requires consideration of the costs and the benefits to society, as well as to the power company, of each option. Production of power from nuclear fission almost inevitably involves production of material from which atomic bombs can be made. While it is impossible to quantify the risks involved in this link, it means that the production of useful power by nuclear reactors increases to some unknown extent the risk of all-out nuclear war. This is one of the sources of uncertainty in the cost-benefit analysis of nuclear power that is not shared by other sources of useful energy. One important facet of meeting the challenge of nuclear power is a foreign policy that impedes the proliferation of nuclear bomb materials and decreases the risk of their falling into irresponsible hands. Unfortunately, since the volume mentioned above was written in 1972, the world has not moved forward in this respect and indeed has moved backward. For this and other reasons, nuclear energy appears to be a less attractive option than it once seemed.

In view of the deterioration of nuclear prospects, it is more meaningful to consider the broader challenge, the challenge to meet the energy needs of society—needs and not necessarily demands—without undue risk of catastrophe or excessive harm to the environment. This involves comparison of the prospective costs and benefits of nuclear energy with those of alternative sources of power. The emphasis in this chapter is on those aspects of the nuclear energy problem most relevant to that comparison.

Nuclear Reactor Development

The early development of nuclear reactors along with the development of the atomic bomb was remarkable for two reasons: the rapidity with which a completely new technology was brought, by great industrial effort, to the stage of practical application; and the extent to which its development was dependent on government backing when the perceived need was great enough to merit a high priority. There may be lessons here for the solution of our present energy problems.

In 1911 Rutherford discovered that each atom contains a nucleus much heavier than the electrons surrounding it, and Bohr immediately inaugurated the understanding of how the electrons circulate around the nucleus to make atoms behave chemically the way they do. In the following two decades it became clear from the carefully measured masses of the nuclei that they were built of building blocks bound together very energetically. This implied that if the building blocks could be combined together differently by taking a nucleus apart or combining two nuclei into one, great energies would be involved. One spoke of the great energy bound within the atom. Specifically, it was learned that there were just two ways in which energy could be released, rather than absorbed, either by splitting a very heavy nucleus to make two medium-weight nuclei or by combining two very light nuclei to make a heavier nucleus, but no one knew how to do either one. As late as 1932 it was discovered that there are two kinds of building blocks, the proton and the neutron. They differ mainly in that the proton has an electric charge and thus interacts strongly with the electrons in any atom that it hits. The proton does not go very far if it is shot into a piece of matter, whereas the neutron has no electric charge and passes easily right through many atoms until it happens to hit the very tiny nucleus of one.

Following this discovery, throughout the thirties, a great deal was learned about nuclei, all done in the spirit of pure science, unlocking the secrets of Nature with practically no consideration of practical applications. In late 1938 a discovery by Hahn and Strassman in Nazi Germany happened to have a tremendous practical application. They found that a uranium nucleus, when hit by a neutron, can split in two. The two fragments fly apart with a remarkable amount of energy that is converted into heat as they slow down in passing through matter. Each of them, when it is stopped, is radioactive. This means that the fragment emits ionizing radiations that can produce more heat or, if they pass through living matter, can cause cell damage with pathological consequences. There are several ways the uranium nucleus can split in two in the fission process, giving rise to quite a variety of fragments or fission products, each with its individual radioactive and chemical properties.

The surprising news of the discovery of the fission process reached the United States and England in early 1939 and it was soon discovered further that the process also emits neutrons. This suddenly made it possible in principle to create a chain reaction: a neutron causes a uranium nucleus to undergo fission that emits neutrons that cause further uranium nuclei to undergo fission and so on, the number of fissions growing larger and larger. In theory these numerous fission processes could release a great deal of energy unless the developing explosive force would blow the mass of uranium nuclei apart before very many of them had undergone fission.

Thus just at the beginning of World War II in September, 1939, it became a serious question whether a very powerful bomb, a hundred or perhaps a thousand times as powerful as any bomb known, could be invented and constructed. Even before direct United States involvement in the war, the most disturbing question was whether such a bomb might be developed secretly in Germany and give Hitler the power to become dictator of the world. It was simply not known whether this was possible but it was felt by American and British scientists that the attempt should be made so that Hitler would not be the sole possessor of such weapons if it could be done.

The remarkable fact is that in the six and a half years following the discovery of the fission process, the necessary scientific investigations were made of the properties of many different nuclei; nuclear reactors were invented, designed and constructed for producing a new fissionable material, plutonium; enormous factories for preparing nuclear materials were created; the mechanisms of the first atomic bombs were invented—and two Japanese cities were destroyed by two bombs, each one a thousand times as powerful as any earlier bomb. All this happened while there were many other heavy wartime demands on American industrial productivity and technical ingenuity. Several other simultaneous wartime achievements including radar, the proximity fuse and artificial rubber, similarly demonstrated the remarkable capability of our industry and technology for very rapid accomplishment once the incentive has been recognized, the decision made, and a high priority established. Given the priority, development of huge-scale wind power in peace time should be simple in comparison.

The first nuclear reactor was a relatively small one built on the campus of the University of Chicago and in it the first fission chain reaction was achieved in late 1941. It used pure graphite as moderator in place of the water used in most modern power reactors because graphite is better than ordinary water at slowing down neutrons without absorbing them. The reason for slowing down the fast neutrons emitted during fission is that uranium consists of two kinds of nuclei, or isotopes, and fission is induced very easily by slow neutrons in the rare isotope called U-235, whereas the abundant one, U-238, undergoes fission with much smaller probability when hit by fast neutrons. Modern commercial reactors can tolerate some loss of neutrons in the water that slows down the neutrons because they use uranium enriched in U-235 (enriched to about 3 percent rather than the 3/4 percent found in nature). This was not available for the first reactor.

The rare isotope, U-235, is also much more susceptible to fission by fast neutrons than is U-238 and uranium highly enriched in U-235 (or else plutonium) is required for making bombs. The enrichment process is difficult and costly. An enormous factory for that purpose was built at Oak Ridge, Tennessee, that produced the material for the Hiroshima bomb. About ten years later two similar plants were built at government expense to produce bomb material and when that demand was largely satisfied their enriched uranium production became available for commercial water-moderated reactors. It is this availability that has made possible the compact design of American commercial reactors. This is one important way in which military and civilian nuclear technology, nuclear bombs and electric power production, are inextricably intertwined. It makes it difficult to calculate a true cost of nuclear electric power.

Largely on the basis of experience with that first small reactor in Chicago, very large reactors were built in Hanford, Washington, also using graphite as moderator, to produce plutonium for atomic bombs such as the one that destroyed Nagasaki. They also produced a lot of heat as a byproduct that was wasted by heating the Columbia River. Because of the wartime haste the reactors were not designed with the accompanying machinery to generate electricity.

After the war it was decided to continue improving and stock-piling atomic bombs and to preserve and expand the nuclear research and development establishments of the wartime effort as peacetime national laboratories. While there was need for plutonium for bombs, this provided the opportunity and talent to develop reactors designed primarily to produce electric power but with plutonium as a byproduct. This was a major and successful effort of Oak Ridge and Argonne national laboratories during the period of 1946 to 1954. The pressurized water reactor (PWR) was developed at Oak Ridge and the boiling water reactor (BWR) at Argonne. When the prototype BWR at Argonne came into operation the five megawatts it generated could not be fed into the commercial grid and was instead used on laboratory grounds because the utility companies were guarding against socialistic government encroachment on their electric power business.

The development of reactor technology was progressing smoothly and efficiently on a reasonable scale in the national laboratories until 1954 when the nuclear authorities, led by the very influential chairman of the Atomic Energy Commission and special advisor to President Eisenhower, Admiral Lewis Strauss, decided that it was time to bring the strength and talent of American industry into the picture. There was public complaint at the time that the negotiations to persuade the Soviet Union to agree to stop the nuclear arms race were not being pursued seriously enough. The very name of the program for rapid expansion of the nuclear power industry "Atoms for Peace," reflects the intention that it should distract attention from the military role of the atom. Except for knowledge directly concerned with weapons and isotopic enrichment, nuclear secrecy was greatly relaxed as part of this program, which incidentally facilitated nuclear applications in medicine and elsewhere, and the valuable results of the reactor development and design work in the national laboratories were handed to industry on a silver platter, so to speak. The government was also in a position to give industry the benefit of the large expenditures it had made to prepare uranium fuels, all the way from prospecting to isotopic enrichment.

Industry was, however, understandably reluctant to embark on so new a venture and it had to be encouraged by some artificially favorable arrangements. One was the Price-Anderson Act of 1957, a remarkable piece of legislation that treats the nuclear industry as no other industry is treated, essentially absolving it from responsibility for the most serious damage that a cataclysmic failure of a nuclear reactor could cause. The details call for private insurance up to the relatively low limit to which insurance companies are willing to assume the risk, then government insurance to a higher limit, and complete irresponsibility beyond that.

Another enticing arrangement was that the government would rent to industry enriched uranium fuel to power the reactors, which would then be returned to the government for reprocessing, while the government would buy from industry the plutonium content of the returned fuel. When these inducements became attractive enough for one utility company to "go nuclear," it was attractive enough for many to do so and the construction of commercial nuclear power plants began with quite a rush. For purposes of financing and regulation they were at first labeled "experimental" nuclear facilities, as indeed they were. Rather than leaving the experimental development to the national laboratories, the consequence was that each "experiment" was carried out many times simultaneously, multiple costly mistakes were made before the development was far enough along, and a large industry was committed to nuclear power before enough was known about the difficulties to make wise judgments. It did have the practical advantage of providing early experience of reactor technology in a commercial environment and it has by now generated substantial amounts of electric power. The very extensive development program has absorbed energies and funds that might have gone into the development of alternative sources of power.

Advantages of Nuclear Power

Compactness of Fuel
The wonderful thing about nuclear power is the compactness of its fuel. The energy of chemical fuels arises from interactions between whole atoms. Nuclear energy arises from interactions within atomic nuclei. Nuclei are about ten thousand times smaller in diameter than atoms and the energies involved in individual nuclear reactions are about ten or a hundred million times as great as those in chemical reactions. If all its nuclei undergo fission, a pound of the uranium isotope U-235 is equivalent to four million pounds (or two thousand tons) of coal. Actually the "burnup" of U-235 in a reactor is only about 10 percent whereas coal burnup is close to 100 percent. Also, the enriched fuel used in United States reactors is only about 3 percent U-235, but even so this enriched uranium fuel is, weight for weight, about ten thousand times as productive as is coal.

When one contemplates the long trainload of coal required to fuel a large coal-fired power plant for a day, this looms as a great advantage. The advantage is somewhat discounted by the fact that uranium fuel cannot be shipped around in loose bulk the way coal can. It requires massive containment. Especially on the way back to the processing plant when the fuel is highly radioactive because of the fission products it contains, the cask in which a fuel assembly is shipped weighs almost a hundred times as much as the fuel itself and the cask must be handled with extreme care to avoid accident. Even with allowance for this extra care, the transportation requirement for uranium is much less than a tenth that for coal.

There is a corresponding advantage in the handling of the smaller amount of fuel within a nuclear power plant, although here again great caution must be exercised to avoid exposing plant personnel to excessive radiation. The plant must be closed down for a few weeks about once a year to replace spent fuel assemblies but aside from that in normal operation the fuel works out of sight, silently taking part in the chain reaction. The normal operation is so seemingly simple, neat, and clean that there is an almost antiseptic impression about the whole plant: the operators sit at the control panel with all its meters and signal lights, the turbogenerator hums away sending electricity over the power lines, and somewhere outside there is a rush of the warm-water discharge or perhaps a gentle plume from the top of a huge cooling tower. That is about all there is to see and no one seems concerned about the fuel that is locked up inside where no one is allowed to go because of the high radiation level. By contrast, a coal-burning plant has all the disarray of great conveyer belts and stokers throwing a continuous stream of dirty coal into the flames. The contrast with oil-burning and natural-gas-burning plants is not so great but, compared with coal, their days are numbered.

Fuel Supply with the Breeder

At times it has been possible to point to the relative cost and expected continuity of uranium supply as an advantage of nuclear power plants but the cost of all fuels has been changing so drastically that this cannot be relied upon as an advantage. In the past, when government preparation of the nuclear fuel has favorably affected its price, the higher capital costs of nuclear plants has about offset the fuel-price advantage. The prospective future supply of uranium cannot be counted as an advantage for nuclear power as we now know it, since the supply of reasonably usable coal is expected to last about ten times as long as the supply of reasonably useful uranium, two hundred or three hundred years as against twenty or thirty.

This situation is of course the motivation for the breeder development program. If the prospective United States type of breeder reactor, more compact and presumably more efficient than the European types already in experimental operation, can be satisfactorily developed, it will have the advantage that its fuel supply will last a long time. It consumes the uranium isotope U-238 which is 140 times as abundant as the rarer isotope U-235 consumed in the present water-moderated reactors and thus extracts about a hundred times as much energy from a lump of uranium ore. This extends the expected lifetime of the global uranium supply by more than a hundred times, eventually after the slow breeding process, because it makes it worthwhile to mine lower grades of ore.

Some of the radioactive fission products that can be separated from the spent reactor fuel of today's reactors have important medical and industrial uses. This, however, cannot be claimed as an advantage of having a large nuclear power program. It would be an advantage of having at least one or two reactors operating in the country, but the amount of these radioisotopes that can be usefully employed is much less than is produced by the many reactors now operating. In fact, most of the radioisotopes that have been used in research have come from one relatively low-power reactor specialized for this purpose at Oak Ridge.

Disadvantages of Nuclear Power

Proliferation and the Risk of Nuclear War
All-out nuclear war is such a horrifying prospect, threatening to devastate completely all we value in life and civilization, that avoiding it is the most important objective of all our public endeavors and decisions. The past thirty years of recent history free of nuclear war have dulled our concern but do not prove it impossible or even unlikely. Avoiding nuclear war at least in the short term is an actual objective of present foreign and military policy dealing with the relationships between the two major nuclear powers. As the number of countries with nuclear power increases, the effectiveness of these efforts to prevent an outbreak of a nuclear war somewhere in the world will fade, and any nuclear war anywhere may escalate into worldwide nuclear war. The prospect of a world with many nuclear-weapons nations is a diplomatic and military nightmare. Given the present trends in the promotion of nuclear power, this seems almost inevitable and the question may be when rather than if. The longer this situation can be delayed the better. A more optimistic view is to consider whether the proliferation of plutonium-making facilities accompanying the spread of nuclear power may be post-

poned long enough that alternative power technologies become attractive and the worldwide use of nuclear power may never reach the plutonium-economy stage. This would buy time that might also allow for diplomatic resolution of some problems.

In assessing the advantages and disadvantages of nuclear power, we should decide whether our present expansion of nuclear power increases or decreases the likelihood of nuclear war. Some may say it does not make much difference. Even if this is true, even a slight alleviation of this risk would be very important. When considering an occurrence as utterly catastrophic as all-out nuclear war, even a small influence on its probability should be a matter of great concern. (The present assessment is close to a limit that might be described mathematically as epsilon times infinity is infinity.)

The conclusion seems inescapable that our present and continued use and promotion of nuclear power substantially increases the risk of nuclear war. But there may be influences the other way and this conclusion is a result of balancing pros and cons. The consideration must include postulating what sort of world demographic and economic development is desirable and achievable (Meadows, 1972; Kahn, 1976; Lovins, 1975; Weinberg, 1970; Inglis, 1971).

The main argument in favor of the spread of nuclear power plants would claim that the only way eventually to check the tendency of the underpriveleged majority of mankind to multiply so prolifically is to industrialize the globe in our image, bringing those rapidly increasing billions closer to our standard of living or at least feeding them adequately and providing enough social security to reduce the economic need for sons. Overlooking how much nuclear power would be required to accomplish this, the argument is that nuclear power would be helpful, reducing the unstabilizing effects of social unrest along the way and thus decreasing the pressure that might lead to nuclear war. This argument of course subsumes that there is no satisfactory alternative to nuclear power. The prospect of developing disparate regions of the globe in our own image is in keeping with economic objectives of promoting foreign trade. This idea, rather than welfare of the people or concerns about nuclear war, may dominate decisions.

The world population problem is vast and historically fraught with human suffering, and is only postponed by medical technology and sanitation, which mitigates pestilence without yet having a compensating influence on birthrates. It seems unlikely that nuclear power or any power source dependent primarily on foreign economic initiative and high technology will make much of a dent in the problem. As long as there is expectation of a miraculous new power source from abroad, even if it is slow in materializing, there will be less pressure for and appreciation of positive measures to limit population growth. Development of power sources should be encouraged, to help where they can, but sources attainable in the long run by local engineering talent and industrial initiative would be healthier for the future of most countries than if they succumb to a neocolonial dependence on foreign high technology.

While the uranium used in most reactors is not sufficiently enriched to be suitable for making nuclear weapons, every nuclear reactor is a producer of plutonium, a different chemical element that can be separated from uranium in unspent fuel by chemical means and used to make nuclear fission bombs. Avoiding exporting fuel reprocessing plants is a useful delaying tactic, making this more difficult for a while, but it cannot be expected to keep the lid on for long. In view of the relative ease of converting plutonium oxide or pure plutonium into bombs, a world with plutonium as a plentiful item of trade is a frightening prospect. Despite whatever precautions can be taken, it seems likely that its presence will hasten the emergence of dozens of nuclear-weapons countries and the time when terrorist groups will have the option of operating with crude atomic bombs (Willrich, 1974). As the number of countries having nuclear weapons rises, so do the chances that a nuclear war could break out and develop into total nuclear war. This increased risk of the ultimate calamity resulting from the availability of plutonium seems to outweigh any stabilizing effect that might be gained from power supplied by nuclear means, particularly since the power could in most cases be supplied by other and more appropriate means. The spread of nuclear power plants throughout the world does indeed increase the risk of nuclear war.

Global attempts to slow down or prevent the proliferation of nuclear materials and their diversion to nuclear weaponry have taken two principal forms, the safeguards program of the International Atomic Energy Agency (IAEA) and the Nonproliferation Treaty (NPT) of 1970. The IAEA safeguards depend largely though not entirely on occasional inspections of nuclear power plants subject to the program. The inspectors do inspect the operation procedures of the plants but depend mainly on access to the plants' bookkeeping in trying to assure themselves that there has been no clandestine diversion of fissionable material. This does act as a deterrent. It makes undetected diversion more difficult but not at all impossible.

The first enlightened and even generous impulse of American nuclear diplomacy at the beginning of the atomic age, before the arms limitation negotiations settled into mere shadowboxing, was based on the realization that having inspectors of a control organization merely look over the shoulders of the nuclear power plant operators could provide no assurance against diversion. At that early stage of the United States monopoly it was proposed in the Acheson-Lillienthal-Baruch Plan that an international agency should be created having sole responsibility and management of all global nuclear development. The real world has not been ready for such an intrusive arrangement and without it there can be no adequate safeguard.

Technical secrets cannot be kept forever. If a small independent group has plutonium, the information necessary to make an atom bomb is increasingly available. Indeed a recent book, motivated by a desire to make authorities take safeguards more seriously, has assembled in one place a general outline of bomb-making know-how (Willrich, 1974). This makes it urgent that the world's plutonium supply be kept in responsible hands and that no black market be permitted to develop. Yet in a world in which drug traffic can be no better controlled than it is, this seems an almost hopeless task as plutonium becomes much more plentiful. One possibility would be to stop the production of plutonium by giving up nuclear power. This precarious situation must be considered a serious disadvantage of having a nuclear power program on a global scale.

The question may be raised, however, whether this is a disadvantage of our going ahead with a very large national nuclear power program (Shelling, 1977). It can be argued that if we do not continue and contribute to nuclear proliferation, other nations will, so what we do does not make much difference. This question is particularly acute in regard to the export of nuclear power plants, which is a natural profit-making accompaniment of a large national nuclear power program. American firms have been world leaders and we are thus largely responsible for the present state of the export business, but if we should simply stop exporting now, Canada, France, and West Germany would continue. It may further be argued that it helps the safeguards situation for us to get as much of the foreign business as possible because the United States requirements for safeguards covering exported plants are more stringent than those of the other exporters. United States law requires that exported plants must be subject to IAEA safeguard provisions but it does not go so far as to limit export only to countries whose other nuclear facilities are subject to IAEA safeguards, as are those of the non-nuclear-weapon countries which adhere to the Nonproliferation Treaty.

United States reactors are dependent on enriched uranium (as Canadian reactors are not) and the fuel is normally returned to the United States for reprocessing. This in itself is an important safeguard in preventing diversion of plutonium in the recipient country, since the fuel-reprocessing plant where plutonium is extracted is the weakest link in the fuel cycle. As a striking example of the cutthroat nature of international competition in this field and the greater concern for safeguards in the United States, a West German firm in 1975 beat out Westinghouse in a bid for export to Brazil largely because the West German package of nuclear facilities covered the entire fuel cycle, including fuel-reprocessing facilities, and United States law would not permit Westinghouse to match this.

While it will be some years before this new Brazilian contract will materialize, this shows how rapidly the situation is deteriorating. In 1974 India became the sixth nation to have tested a nuclear explosive, surreptitiously using nuclear material from a reactor supplied by Canada in defiance of its word to Canada. Now it appears that Brazil may go further.

Even with the proliferation prospects as serious as they are, it will make a difference whether the United States continues its expansive nuclear power policy. The sheer volume of the plutonium economy will exacerbate the problem and the United States has the largest industrial capacity in the world to contribute to that volume.

It has been proposed that the United States could delay the advent of the plutonium economy, especially in view of the development of alternative energy sources, by continuing the present state of affairs in which nuclear fuel is used on a once-through basis, omitting the fuel-reprocessing part of the former fuel cycle (Feiveson, 1976). Surprisingly, this practice, which was initiated out of necessity for lack of operating processing plants, seems to be economic or nearly so. That is, it barely pays to process the fuel for the sake of recovering the residual uranium-235 in it. This would mean reactors exported from the United States or from Europe under American license would be supplied with fresh fuel elements mostly from this country, the spent fuel elements either to be returned here or stored locally without reprocessing so that the plutonium would remain locked in an intensely radioactive solid mixture not easily accessible for nefarious purposes. Thus there would be no traffic in easily convertible plutonium. This arrangement could be continued only as long as the economic supply of fresh uranium lasts, then there would be economic pressure to separate out and use the plutonium. Thus the longer the supply of fresh uranium lasts, the longer the dangers of the plutonium traffic can be post-poned. In view of the practical appeal of this policy option, which is indeed a part of President Carter's energy proposals as is dis-cussed in chapter 11, it can be said that each nuclear plant in opera-tion, by hastening the end of the supply of fresh uranium, is adding slightly to the risk of nuclear war.

As is discussed further in the last chapter (chapter 11), it will also make a difference whether the United States continues its pri-mary emphasis on nuclear development or shifts priority to the de-velopment of alternative sources of power better suited to the more modest technical talents of the developing countries, sources that could compete against nuclear export from other countries as well as our own. On both counts, then, the threat of nuclear prolifera-tion and its influence on the likelihood of nuclear war must be counted as a disadvantage of continuing our present emphasis on nuclear power.

Despite the remoteness of the possibility that any enemy would dare attack the United States and in spite of our natural desire to dismiss the idea as unthinkable, we spend many billions of dollars annually on retaliatory capabilities, presumably to discourage such aggression. Another way to discourage it is to reduce the vulnera-bility of our civilization to attack. Centralized power plants make tempting targets. Knocking them out by a selective strike would bring the entire country to a halt. Their destruction would decrease the chance for some organized activity to rise from the ashes of nuclear war. Nuclear power plants make especially inviting targets because, if one were vaporized by being within the fireball of a nuclear weapon, the fission radioactivity from the weapon itself would be enhanced a thousandfold by the vaporized contents of the reactor core. Grotesque as it may sound, even this contribution to the likelihood of nuclear war must be counted as one of the dis-advantages of nuclear power especially when compared with wind power from many thousand wind dynamos half a mile apart on the Great Plains.

Radioactive Releases from Usual Operations
The amount of fission-product radioactivity in an operating power reactor is so enormous that a very high degree of perfection is re-quired to keep it contained. In normal operation of the reactor this can be attained with sufficient care and normal releases are required to be kept at a very low level as a result of recent public pressure. It has not been found possible to keep the releases from fuel pro-cessing plants quite so low in normal operation and there are fre-

quently abnormal emissions, called "burps," from the reactors themselves. The incidence of these emissions raises questions about possible pathological effects to the public, and merits discussion here. These releases, although stronger than normal emissions from reactors, are still low-level radiation. It is difficult to obtain significant statistics on the effects of low-level radiation because they are distributed thinly throughout a large populace subject to other pathological influences. The same difficulty is encountered, though under better control, in the study of laboratory animals.

One way to bypass this difficulty is to make the usual reasonable assumption that pathological damage, both genetic and somatic, is proportional to the radiation dose received, even if accumulated over a long period. The effects of large doses are known from the experiences of the unfortunate survivors of the two bombed Japanese cities and from a few accidental exposures. From this information, one can extrapolate down to very low doses. According to this assumption, all radiation is harmful, but the lower the dose, the less the harm.

The unit of radiation exposure is called the "rem." There is a natural background radiation coming from ordinary rocks and from outer space as cosmic rays, amounting on the average to about one-sixth of a rem per year of exposure. People get on the average about half as much as this in addition from medical X-rays. In taking or prescribing X-rays, one should weigh the expected benefits against the possible harm.

A rather arbitrary limit is commonly set on the radiation dosage considered tolerable. As more has been learned about radiation pathology through the years, the limits have been steadily reduced. In 1970 it was considered permissible for nuclear activities to increase the exposure to the public by one-sixth rem per year, adding the amount of the natural background to the natural background. The rule for nuclear power plants then was that their emissions should not exceed that level at the plant boundary, where people might live near the plant.

It is technically possible to keep levels much lower than this but it costs money to do so and there was resistance to setting the permissible level lower. Some felt, however, that this was unnecessarily high and two scientists, Drs. J. W. Gofman and A. R. Templin urged that the limit be lowered in keeping with their estimate that, if the entire population of the United States were to receive this "permitted" dosage, it would cause about thirty thousand cases of cancer per year and some similar large number of genetic defects. Of course the population does not live at plant boundaries but it can further be estimated that if there were a thousand nuclear plants in the country each reaching this limit at the plant boundaries, the average exposure would be at one-tenth this level. This would mean about one to three thousand extra cancer deaths per year, or, roughly one per plant per year.

Shortly after this discussion, in 1971, the AEC agreed that the limit should be lowered and set the "guideline figure" at one-thirtieth of the former limit (namely, at five millirems per year), stating that future licenses would be issued only to new plants planning to stay within this new limit. While relatively few plants pushed the old limit, it seems likely that normal operations may be not far under the new limit so that one cancer death per thirty or per hundred plants per year might be an actual figure. It is gratifying that this figure has now been made so low and need no longer be considered a very serious disadvantage of nuclear power.

Similar new restrictions have not been applied to fuel processing plants because, when they were still operating, it had not yet been found feasible to keep their routine emissions well below the older limits. Their releases take the form of both gases and liquids. One way to keep emission low at power plants is to keep the radioactive gases contained either within the fuel elements or in bottles. Most of the gases are released later at the fuel processing plant. In fuel processing almost all of the highly radioactive liquid and dissolved solid waste is concentrated into compact form for storage but some of it appears in diluted form as wash water, which is normally flushed into drainage. The radioactivity in the creek draining the West Valley, New York, fuel processing plant was once found by outside investigators to be at several thousand times the permitted level. That plant has been shut down and construction of another discontinued in accordance with the 1977 proposal to avoid reprocessing for the present. This will delay the advent of the plutonium economy but it is not clear how long such a decision may be in effect. Without more information on future plants it is hard to say how serious an objection to nuclear power this source of contamination may be. It is complicated by the fact that radioactivity is concentrated in food chains, increasing in intensity from water to plankton to small fish that eat the plankton to big fish that eat the small fish, for example.

One of the conventional claims for nuclear safety is that there has never been a fatality in the public due to operation of United States commercial nuclear power plants. If one modifies this statement to say "clearly identifiable as due" to these plants, this is true. Though statistically one expects that there has been a very slight increase in incidence of cancer deaths due to nuclear power plants, as just mentioned, any individual cases are lost in the general incidence and remain unidentified.

Another approach to the problem is to assemble vital statistics on infant mortality and to try to disentangle the effects of possible local low-level radiation exposure from other environmental and sociological effects, despite the difficulty of obtaining adequate statistics. The principal practitioner of this method, Professor E. J. Sternglass, has remained a controversial figure partly because of his unrestrained enthusiasm in advocating his findings which are at some variance with those obtained by other methods. This should not be allowed to foreclose serious consideration of this aspect of the radiation problem.

Radiation damages individual cells within the human body and thereby influences the process by which the cells reproduce themselves. For this reason rapidly growing tissue is most susceptible to radiation damage, especially in the unborn fetus and in newly born infants. Both somatic damage, which affects infant mortality, and genetic damage, which is more difficult to find in vital statistics since it may not show up until later, are important.

The first time any of the United States population was exposed to the low-level radiation from an atomic bomb test provides especially significant data because the population had not been previously so affected. The test took place in southern New Mexico in 1945. Weather bureau reports showed that winds at intermediate altitudes were to the north but the mushroom cloud at higher altitudes was seen to move off toward the east. The infant mortality the following year in a swath of states reaching as far east as North Carolina is reported to have revealed an apparently significant rise involving large enough numbers to be statistically valid.

Observation of infant mortality in the counties or towns near nuclear plants involves smaller numbers of cases and remains statistically more questionable, but the persistence of many suspicious situations can be disturbing and is in need of further study that would cover as many cases as possible. In counties near plants that have released radiation into the atmosphere, infant mortality has apparently increased in some counties but not in others and furthermore, does not show good correlation with prevailing wind direction. When a reactor in Pennsylvania had a period of abnormally high liquid release into the river, downstream towns using the river for drinking water were found to show a significant increase in infant mortality, which tapered off farther downstream.

On the basis of such data, Sternglass claims that the number of deaths due to low-level radiation from nuclear power plants is between ten and a hundred times as great as surmised from the figures quoted above. Those figures are based partly on the assumption that the mechanism of radiation pathology is damage to the genetic code in the reproductive mechanism of the cell. The discrepancy in the figures suggests that there must be another mechanism operating and indeed there is laboratory evidence that the functioning of cell membranes, on which the life of a cell depends, is adversely affected by radiation. This might be the source of some of the apparent excess in infant mortality. If this were the principal mechanism, infant mortality would not be accompanied by proportionate genetic damage.

There is much uncertainty in the numbers encountered in such discussions of low-level radiation pathology. It is a matter of judgment whether one should be concerned with adverse effects suggested but not definitely established by the available statistics. One attitude is that we should not interrupt important practical technical and commercial progress because of concern for possibilities that cannot be proved to exist. A more cautious attitude is that the very possibility that a program is doing mortal harm, even if it cannot be established as proven fact, should be weighed in the balance of costs versus benefits of the program. According to the latter view, the risk of appreciable infant mortality and malformation is already a disadvantage of a nuclear power program.

With cancer deaths the situation is somewhat different. It is generally admitted as fact that there will be some cancer deaths induced by the nuclear power program, but they are an extremely small percentage of the tragic death toll due to cancer, perhaps a ten-thousandth of a percent at the lowered permissible levels of emission for power plants and still a very small percentage for the entire fuel cycle, even if we include the cancer of uranium miners.

A familiar claim for nuclear power is that the pathological effects of emissions are an argument in favor of nuclear power because in this respect coal-fired plants are more deadly, with their sulfurous and other emissions. Also, there are many more coal miners suffering from black lung disease than uranium miners contracting cancer. Insofar as coal is considered the alternative, as is common, this is valid. In its more plentiful grades, coal is dirty in more ways than one and, no matter what supplements to coal are developed, we will be burning coal for some time. But here the argument is not between coal and nuclear fission, but between wind and nuclear fission as supplements to coal. In this comparison, radioactive emissions are a distinct disadvantage of nuclear power. One of the attractive features of solar power in general and wind power in particular is cleanliness.

Risk of Calamitous Accident

Nuclear reactors in nuclear power plants are designed with the realization that they are potentially extremely dangerous and that it is absolutely necessary to design them to contain their radioactive contents with essentially complete certainty. The danger arises from the fact that a typical reactor in normal operation contains about a thousand times as much long-lived fission-product radioactivity as is produced by an atom bomb of the size used against Japan. If as much as a thousandth of this radioactive content of a reactor, or one atom bomb's worth were to escape, it could do far more damage than does the radioactive debris of an atom bomb because the reactor release would remain close to the ground whereas radiation from an atom bomb is carried aloft by the tremendous heat of the blast to form a mushroom cloud. After much of the radioactivity has decayed, the remainder slowly trickles back to earth widely dispersed.

Many features of the design of a power reactor are dictated by necessary safety precautions. There are normally three barriers, three lines of defense in depth, preventing the escape of fission products from the fuel pellets into the atmosphere. The thin cladding of the fuel rods is snug around the pellets. If the cladding leaks, as it may through pinholes that develop, some radioactivity gets into the circulating water, which is contained in the very thick stainless steel reactor vessel that maintains the high pressure of the cooling water. Both of these lines of defense are necessary to the functioning of the reactor, but the third line of defense is purely a safety precaution. In one type of reactor it is an outer containment consisting of a steel and concrete pressure vessel with a hemispherical dome, capable of containing for several hours the contents of the main pressure vessel if they should be dispersed into the larger volume. Within this outer volume there are arrangements, usually involving water sprays, for removing heat to prevent excessive build up of pressure.

The precautions for assuring the capability of shutting down the reactor in emergency are even more elaborate. A reactor is turned off or "scrammed" by insertion of neutron-absorbing control rods and these must function very dependably. They receive their commands from electronic circuits and in order to assure that failure of one component will not block the functioning, there are several circuits simultaneously passing on the command, so that if one or even two fail, others will perform. This shows the redundancy that is widely used to assure proper operation of the controls.

One reason it is so important for a reactor to shut down dependably and automatically is that a power station sometimes suddenly loses its load. This can happen several times a year, when lightning strikes a power line and opens a circuit breaker, for example. With no place for its power to go, a reactor would start to melt in a few seconds if not shut off. But even if it is shut off by stopping the chain reaction it continues to produce about one-tenth as much power as before because of the radioactivity of the fission products. This heat is enough to start to melt the reactor core in half a minute. To remove this heat in case of main circulation failure, an emergency core-cooling system is provided that must turn on automatically. Unfortunately, there are serious doubts as to whether this system will work in boiling water reactors. Some preliminary small-scale tests have unexpectedly failed and full-scale tests are not practicable. The reactors continue in use, nevertheless, since it is considered unlikely that both the load and the main circulation will be lost at once.

With a large number of precautions such as these, engineers have succeeded in making it unlikely that any given reactor will have an accident with extremely serious consequences in any given year. Yet we must consider both what the consequences might be and how unlikely they might be.

One subject of some concern is a possibility of chemical explosion following a melting of some of the fuel. If water comes into contact with a very hot molten mass, it could be dissociated into oxygen and hydrogen, which might explode, perhaps ejecting fragments of the main pressure vessel as projectiles that might pierce the outer containment. Zirconium alloys are used as a cladding material to contain fuel pellets because of favorable properties under intense neutron bombardment; yet chemically zirconium is one of the most reactive metals and when in contact with water at high temperature is another source of hazardous hydrogen (Bulbransen, 1975). Conventionally this is considered extremely unlikely but it suggests the kind of bizarre circumstances one wants to be sure to prevent. But the consequence of a core meltdown that causes the most concern is the expectation that the molten mass would melt its way through the bottom of the reactor vessel and outer containment into the earth beneath (or into the sea if the new idea of floating reactors is in practice). This is perhaps the most likely way that a substantial fraction of the fission products, including especially the gaseous ones, could reach the atmosphere.

Since the potential danger of nuclear power plants is so great, extensive efforts have of course been devoted to ensuring their safety in the course of their development and promotion. There have been two main thrusts of these efforts, in the development aspect to make them as safe as is possible with reasonable effort, and in the promotion to convince both the buyers of nuclear plants and the general public that they are safe. While reactors are safer than they would be without the prolonged devotion to safety development in the national laboratories and the elaborate design and operational precautions required of industry by the regulatory agency, the interactions of the promotional and the developmental efforts (particularly the overzealousness of the promotion) have undermined the credibility of official statements, making it difficult to judge the actual state of reactor safety.

An important part of the government effort to assess the state of reactor safety without raising undue alarm has been a sequence of three formal safety studies spanning the years 1957 to 1975. The first and third studies led to public reports (WASH-740 and WASH-1400) but the second, in 1965, was so pessimistic about safety that it was not reported (until revealed much later after litigation). The first and second attempted only the assess the consequences of an accidental massive release of radioactivity. The first study concluded that if a relatively small reactor eighteen miles from a city were to release half of its radioactive contents into the atmosphere it might kill about three thousand people and do $7 billion worth of property damage. The third study also made elaborate estimates of the absolute probability of the occurrence of various types of accidents even though some participants of the second study and some outside critics consider it virtually impossible to do so reliably, since there are many intricate ways for things to go disastrously wrong in so complex a mechanism as a nuclear reactor.

The methodology of the studies is useful in eliminating some accident sequences from consideration, but there are others that have not been or cannot be assessed and perhaps still others that have not even been imagined, as is suggested by severe accidents that barely avoided becoming calamities, which were caused by failure sequences that had not been anticipated. The third study estimates that an accident involving the molten core of a reactor melting its way through the bottom of the containment might occur once in about twenty thousand reactor years. For the least severe accident considered, radioactivity would be released so gradually after filtering through the soil and would be so dispersed in the atmosphere that it is estimated that practically no deaths or serious illnesses would result. The most severe accident considered in the study is estimated to be extremely unlikely: a rupture of the outer containment from a steam explosion when the molten core, an hour or more after shutdown of the chain reaction, penetrates the inner containment and falls into water. The estimated consequences predict over three thousand early deaths, forty thousand early illnesses, and as many latent cancer deaths. An accident initiated by a burst of the main pressure vessel would be expected to exact a greater toll because the chain reaction would be producing power up to the time of the release and more intense radioactivity would escape, but such an accident is considered too unlikely to contribute to the overall risk.

The mildness of these risk estimates has been severely challenged by some of the outside groups of scientists and others who have made critical studies of the third official study after it was released in draft form. Arguments have been given for increasing the estimated casualties by factors like twenty or forty and for increasing accident probability by large factors also. The specific arguments about this controversial subject are too involed to be reported here, though an outline of some of them is presented in appendix 3.

There is a great deal of uncertainty in trying to estimate the probability and degree of severity of accident occurrence on the basis of theory, in view of our inadequate experimental base. Short of establishing some credibility of those estimates, the only firm basis we have for judging the safety of the large commercial reactors is their safety record to date. There have been about two hundred reactor years of experience with all United States commercial reactors, or only about one hundred with the large ones whose safety record is more relevant to future expectations. On the positive side of the record, there has been no calamity, even no identifiable radiation-caused fatality, in the operation of these plants. That is a record to be proud of, and the nuclear industry is proud of it, spending many thousands of dollars advertising it, although without the qualification ''identifiable.'' Reassuring as it seems, all this record really means is that the probability of a catastrophic accident is less than one per one hundred reactor years, $10^{-2}/RY$. The advertisements sometimes mention a thousand or two thousand reactor years of experience, but this includes foreign reactors and small research reactors, which in general have much lower power densities in the core, and submarine reactors, which are built and operated to more exacting standards. Of the submarine reactors it is only firmly known that not more than one has had a calamitous reactor accident, since one United States nuclear submarine was lost for an unknown cause, another for a known non-nuclear cause.

Even if we include this larger number of reactors in the formula, this safety record means only that the probability is less than about 10^{23} per reactor year, or, for a thousand reactors in the future, about one incident of radioactivity escaping from a molten core into the atmosphere per year. This is clearly unacceptable.

One could pare down these consequences if one considers filtration through the soil when the core melts through the bottom, and how consequent damage would depend on the weather. However, the estimate of 10^{23} per reactor year applies also to more sudden and serious ruptures of containment for which these considerations would have less effect. Even though the fact that these substantial power plants have performed intermittently for several years without any apparent radiation-induced fatality may be made to sound very reassuring, this estimate shows that one should not be lulled into a false sense of complacency. The theoretical arguments can do better than that estimate, but how much better is uncertain as can be appreciated from the remarks in appendix 3 or further pursuit of the various reactor safety studies.

This uncertainty about the capability of extensive safety precautions to prevent a disastrous reactor accident must be counted as one of the serious disadvantages of nuclear power.

Malfunctions and Minor Accidents
A nuclear power reactor is so complex that there are many possibilities for things to go wrong. The consequences of things going very wrong, or of several failures at once, may be so grave that elaborate precautions are provided to try to assure that the worst consequences of malfunctions will not be realized, and this itself adds to the complexity and the number of devices that can fail. Even though most malfunctions do not have dangerous consequences, many of them require that the plant be shut down for repair, which can be very costly. Troubles with components in the high radiation zone near the reactor core can be particularly costly because of the great difficulty of making repairs there. In such cases many repairmen are sent in succession to do a job, each receiving a moderate but "permissable" dose of radiation to avoid having one man stay long enough to finish the job but receive an excessive dose.

In the forty reactors operating for part or all of 1973, the latest year for which the report is available, the AEC listed more than 800 reported "unscheduled events," as malfunctions and minor accidents are called. Some were due to human error, some to equipment failure, and some to both. They include a wide variety of malfunctions, from excessive leaks in the fuel cladding, which required shutting down for refueling; to an operator throwing the wrong switch; or relays sticking and interfering with the reactor functions. It was anticipated from the beginning that there would be occasional malfunctions but such a large number of them is unexpected and distressing. It seems to reflect the difficulty of enforcing quality control, despite great efforts, and the problem of recruiting competent trained personnel for routine jobs. Infallible humans are hard to come by.

These surprisingly frequent "unscheduled events" are mostly responsible for the poor performance records of nuclear power plants. Whereas the sales pitch anticipated that the new large power reactors would deliver an average of 80 percent of capacity or better, the actual figure has hovered around 55 percent. As the accompanying graph presented by D. D. Comey (Comey, 1975) dramatically shows, the figure tends to be low for a new plant as start-up troubles are being cured, then to rise to a maximum at a plant age of about four years, and after that it declines ominously, suggestive of early old age. Of course industry hopes that this is a mere fluctuation and that the graph will resume its upward course toward the anticipated 80 percent as more troubles are cured and the technology matures, but as of 1977 there seems to be no justification for this hope.

Radioactive Waste Disposal

At the fuel processing plant, when one is in operation, the spent fuel pellets are dissolved in acid, the uranium and plutonium are separated out for further use and the burden of hot radioactive fission products is separated as waste. Most of the waste is in the form of a concentrated liquid solution that must somehow be disposed of permanently without contaminating the environment. The radioactivity is so intense that even as it decays it will remain potentially dangerous for centuries. For the first few years it is so intense that it is not only very "hot" radioactively but also thermally. It is so hot that the liquid boils continually and special cooling must be provided to prevent dispersal by spattering.

The waste from extensive military production of plutonium and the less voluminous waste from civilian power production (before reprocessing was recently discontinued) has been and is still stored in huge vats, holding about a million gallons each, mostly at the government's nuclear facility at Hanford, Washington. These vats are cooled and stirred mechanically in a program of "perpetual care" that includes transferring the liquid to new tanks as the old tanks start to leak from corrosion after twenty years or so. The problem of caring for these wastes, nearly 200 million gallons of them, is already with us and it is only because the contemplated expansion of nuclear programs makes the waste problem worse that this is a disadvantage of going ahead with nuclear power.

Plans have been developed for handling the waste in a more satisfactory way by reducing the high-level liquid wastes to solid form, to a kind of glass or a ceramic that can stand the high temperatures induced by the radioactivity. Until now this has not been done on a large scale because of expense but it is planned for the future. The cost of this process has been estimated as only about 0.1 percent of the cost of nuclear power.

Even in solid form, the wastes must still be stored to permit cooling and must be completely sequestered from the environment. Their final resting place has not been selected, but presumably will be deep underground. Salt mines have been considered as repositories for both liquid and solid wastes, using the melting of salt to absorb the heat. Where they have not been disturbed by man, the existence of large salt deposits is evidence of the lack of ground water. Since oil and salt deposits are related geologically, difficulty has been encountered in locating large salt deposits that have not been disturbed by drilling, which might permit water seepage.

At present all waste storage facilities are considered temporary. The prevalent view is that it is satisfactory to go on producing these potentially dangerous wastes and storing them temporarily, for a hundred years if necessary as Congressman Hollifield said in a hearing, trusting that technology will by then find a better permanent solution. With the problem already thirty years old, the question of whether we are justified in trying to reap benefits from nuclear power now and leaving this troublesome legacy for future generations may be considered another disadvantage of nuclear power. The disadvantage is not so much the danger to future generations, for it seems likely that one of the methods now being considered will be satisfactory and present no threat if the sequestering is carefully carried out. The disadvantage is rather the nuisance and cost of having to do the sequestering, and perhaps the nuisance and cost to future generations having to assure that the repositories are not disturbed. This is one more way that nuclear industry depends on the government. It seems to require a continuity of political institutions—eventually worldwide—that has not been common in the past.

Thermal Effluents

Heat disposal is a problem that nuclear power plants have in common with other fuel-consuming plants, but it is more severe in the nuclear plants. The reason is that special materials must be used in the nuclear plant to be compatible with neutron fluxes and also that neutron radiation enhances corrosion so that a nuclear "furnace" cannot operate at as high a temperature as a fossil-fuel-fired one. Any of these heat engines operate with a heat input (to the boiler) at a high temperature, and a heat sink or output (that of the condenser and cooling water) at a low temperature. This was discussed in chapter 5.

There is a thermodynamic principle (associated with the name Carnot) to the effect that, given an amount of heat flowing into cooling water at a given temperature, the higher the input temperature the greater is the potential useful power output. In other words, when the cooling water temperature is determined by the river water available, the higher the furnace temperature is, the less the river is heated per kilowatt of power plant production. One result is that the thermal efficiency of modern coal-fired plants is around 40 percent compared with about 33 percent for nuclear plants. The nuclear plant heats the river correspondingly more than does the coal-fired plant of the same power. The problem is with us in either case, but amplified with nuclear power.

One trouble with heating a river even locally is that it may prevent migration of certain desirable fish past the plant. This may be tolerable if a single plant is heating a moderate-sized river only a very few degrees locally, but as several plants are built on the same river they will heat the entire length of the river and especially its estuary. Such a situation threatens the biota. Preserving the natural temperatures of estuaries is particularly important because they serve as an essential breeding ground for much of the marine life of the open sea. The trend is to require plants to use cooling towers, which of course releases the burden into the atmosphere, in the form of heat, and in most cases, additional moisture.

The atmosphere is so vast that from our worm's-eye view it seems like an ideal repository for heat waste and other refuse. Indeed man has been using it as a trash dump on a grand scale since the industrial revolution. From a global point of view the atmosphere is a thin skin of gas hugging the surface of our planet and circulating in a complicated way. It is subject to several delicate balances. As the industrial age progresses there is a serious question of the heat and other effluents from man's increasing use of power upsetting the global balance and drastically altering the earth's environment.

The climate has always been changing and glacial ages have come and gone. In the natural course of the next glacial cycle the densely populated coastal plains will be flooded and man's habitat will be pushed southward by an advancing glacier. These events are not due for about five thousand years, but the schedule could be advanced considerably by artifically accelerated melting of arctic ice floes. However, it seems unlikely that the industrial activities of man could compress the change into as little as a few hundred years. Thus so drastic a change does not seem to be a matter of concern in behalf of our grandchildren or for a few generations beyond.

Nevertheless, less drastic changes are a matter of more direct concern. There could be a change in the precipitation pattern, bringing rainfall to some regions and drought to others, without any great change in the global average. This local rainfall distribution is influenced by the jet stream that circulates around the earth at high altitudes meandering between the polar and subtropical regions on the way. Its snakelike pattern tends to be "locked in" at some longitudes more than others. That is, a northerly section of the stream may hover over the Pacific Coast for a while, then over the Rockies, for example. Just why is not completely understood, but the balance is apparently delicate enough to be considerably influenced by the pattern of heating in the lower atmosphere. If mankind goes on releasing energy by burning fuel at an exponentially increasing rate, with a doubling time of approximately thirty years as over the last several centuries, this will mean the release of energy in the next thirty years will be about equal to the total of that released in all of past history. Therefore the fact that these activities do not seem to have changed global weather patterns very much does not prove that they will not in the future. On a local scale, the release of moisture from the cooling towers of a nuclear power plant, and especially from several in a nuclear park, could change local weather considerably, increasing fog, for example (Abelson, 1977).

The possibility of weather modification is not specific to nuclear power. It is a disadvantage of producing increasing amounts of power by consumption of any type of fuel. The waste heat problem is somewhat more serious with nuclear power, but the chemical effect of effluents is more serious with fossil-fuel consumption. The most severe long-term effect of burning fossil fuels is to increase the carbon dioxide content of the atmosphere, which thereby inhibits the escape of infrared radiation from the earth (through the "infrared window" of the absorption spectrum). This steps up the greenhouse effect and causes overall heating of the global atmosphere.

Thus possible weather alteration cannot be counted as a disadvantage of nuclear power relative to coal burning, but it is a disadvantage for nuclear power when compared with wind power or other solar-related sources. Wind power does not introduce extra energy into the atmosphere, as nuclear power does, but rather extracts mechanical energy and reintroduces it as heat energy, since all electric energy generated eventually ends up as heat in the atmosphere and bodies of water, whether the electricity powers a lamp bulb, a stove, or a motor to drive machinery and overcome friction. This transfer of the form and place of energy could conceivably cause weather changes but it seems unlikely that they could ever be as severe as those induced by a net increase of the energy content of the atmosphere.

Thus the problem of heat disposal is an argument in favor both of energy conservation and of wind power as compared with nuclear power. Incidentally, as mentioned in chapter 5, it is also an argument for wind power or for direct solar generation of electricity (by photovoltaic cells) as compared with solar thermal generation of electricity (by steam turbines) because the latter requires a heat sink lacking in arid parts of the country where the sun shines brightest and water is scarce.

One way to alleviate the waste heat problem is to find uses for heat delivered at a rather low temperature, such as domestic space heating. The normal thermal effluent from a power plant is probably not quite warm enough for this. In order to use this "total energy" concept it is expected that the condenser at the cool end of the turbine-steam cycle would be run at slightly higher temperatures than usual, thus reducing the electric output of the plant for the sake of having a useful heat effluent. With a fossil-fuel-fired plant located near to a city, circulating hot water to residences and factories for space heating would be practical. In a sense, the space heating takes the place of a cooling tower as a means of transferring the heat to the atmosphere, for this is what happens to any heat used to warm a house as it leaks through walls and windows. It does not, however, as in the case of a cooling tower, put additional heat into the atmosphere, as the house would be heated anyway and this operation simply saves burning fuel to do the job. This is not such an attractive possibility for nuclear power plants, especially if they are clustered in nuclear parks far from cities for safety reasons, because there would not be enough nearby houses to heat. Industries to use such heat could be developed. Vegetable greenhouses are a possibility, but it would take a lot of them to use the available heat. Space cooling, on the principle of a gas flame refrigerator, might be too costly with such low-temperature heat. Heat has been piped over twenty miles from a Soviet power plant to a city, but this practice is not expected to be economic here.

Decommissioning of Spent Reactors

It is expected that the useful life of a nuclear power plant will be about thirty years. By then chemical corrosion enhanced by radiation will have taken its toll and some components of the reactor will be beyond repair. All the internal parts of the reactor near the core and the core itself will be so radioactive that they cannot be handled, aside from the routine removal of fuel elements. An ordinary fossil-fuel-burning plant at a similar stage of uselessness would be dismantled and its parts salvaged, at least for scrap metal. Because of the intense radioactivity of the inner parts, this cannot be done so easily with a nuclear plant. Disassembly and sequestering by remotely controlled robots would probably be prohibitively expensive. The necrology of reactors has received very little attention or planning and no provision is made for it in figuring the costs of nuclear power. It seems likely that the reactors and their housings will remain as monuments to the energy appetite of a prosperous earlier generation, occupying some carefully selected choice real estate.

This is one important reason why reactors should be sited deep underground. Why should one create this radioactive nuisance above ground and later wish there were some way to bury it when it could be built underground in the first place? At the end of the useful life of an underground reactor it would be a simple matter to seal up the ground-level openings and leave it where it is out of sight and causing no trouble.

Two other reasons for siting reactors underground are: (1) that it would provide much stronger containment of a run-away reactor accident, taking the place of the usual above-ground outer containment vessel that is designed to withstand not more than three atmospheres of pressure; and (2) it would make effective sabotage or attack, which could disperse radioactivity, more difficult.

Many hydroelectric installations in this country have their large electric generators installed underground in large cavities of solid rock. That it is practical to operate nuclear power stations underground is attested by the fact that there are four such stations in Europe. The largest one at Ardennes, France, has an electric generating capacity of 266 megawatts, about a quarter as great as the largest nuclear power plants.

Ever since 1957 the AEC has been urged by members of the Argonne National Laboratory including the author (Inglis, 1957) among others, to require that power reactors be put underground. Estimates then calculated that the additional cost would be about 5 percent and more recent estimates made in AEC studies and elsewhere (Rogers, 1971; Watson, 1972) have been under 10 percent. While there is some concern about making some of the nuclear operation more difficult, the real reason for not going underground seems to have been the additional cost, which would interfere with the goal of making nuclear power economically competitive.

Since future reactors are still being planned for siting above ground, the decommissioning problem remains a distinct disadvantage of nuclear power.

Present Status of Nuclear Power

As of 1977 there were about sixty nuclear power plants in the United States licensed and at least operating intermittently, making the installed electric generating capacity about 8 percent nuclear. Expansion expectations for the immediate future were represented by about one hundred fifty plants under construction or on order. Of these about half have been postponed or canceled because of an accumulation of troubles; mainly mechanical failure, rapidly increasing costs, and uncertainties concerning fuel, also to some extent because of improved regulation and litigation over environmental effects. The more frequent than expected malfunctions and minor accidents, discussed above, have contributed significantly to the malaise of nuclear power, helping to reduce the average plant factor of the large nuclear plants from the expected 80 percent to nearer 50 percent.

The rapidly increasing rate at which nuclear plant manufacturers were taking orders in the early 1970s has suddenly come to a halt. One of the five companies, General Atomic, has withdrawn from the market and ceased to produce its unique product, the High Temperature Gas Reactor (HTGR). Incidentally, this reactor seemed to have some advantages over others in terms of safety and thermal efficiency but was developed late and few were sold. Another blow to the industry came in 1975; due to quadrupling of the price of uranium, Westinghouse canceled its contracts to supply uranium fuel at a guaranteed price to the utility companies that had bought its pressurized water reactors. Dependability of economic fuel supply had of course been one of the strong selling points of nuclear power plants and this action must discourage future sales. There were only five nuclear reactors sold in the United States in 1975—about a hundred were sold in the preceding three years.

There is a feeling in some government circles that the utility companies' doubts about the future profitability of nuclear power should not be permitted to halt rapid progress toward national dependence on this energy source (solar alternatives having been written off as being for the next century). It is proposed that there should be new commitments for 200 large nuclear power plants to be in operation by 1990, clustered in nuclear parks that include fuel processing facilities, and that the government should provide the subsidies needed to accomplish this goal. Since utility companies are finding it difficult to raise investment capital because of recent doubts about nuclear power, it is even proposed that the government should buy the plants, contracting for their construction, and lease them to utility companies. One cannot predict with confidence whether this is part of the future of nuclear power, along with successful development and general use of breeder reactors. It would be a mistake to underestimate the lobbying power of a large industry that would like to see such a future; yet appreciation of the nuclear power drawbacks is growing and it seems unlikely that Congress will go so far in spending the people's money for further support of nuclear power.

The fuel reprocessing situation is in the throes of transition between government and private industrial operation. Since 1972, when the privately owned West Valley, New York, plant was closed for rebuilding after numerous troubles, there has been no facility for reprocessing the spent fuel of United States power reactors. Instead of being reprocessed, the spent fuel has been piling up in various temporary storage facilities. A normal part of a nuclear power plant is a storage facility, one or more large tanks of water appropriately arranged under a crane, where spent fuel, after being removed from the reactor by remote control, is normally stored for the six months that it is radioactively "hottest," to let it cool off before shipment to a reprocessing plant. Extra storage capacity is usually provided as a safety measure, to allow for removal of all of the fuel of a reactor core in case of trouble. Instead, in 1977 the spent fuel from recent years' operation was being stored there pending completion of other handling facilities and some plants faced the necessity of closing down unless additional storage facilities could be arranged.

This awkward and hazardous situation arose because a planned fuel processing plant failed to come into operation. It was being built by General Electric at Morris, Illinois, employing a modified process that was expected to save costs. After spending $55 million on it (a small sum in the nuclear energy business) the company abandoned the enterprise as impractical. A new fuel processing plant to be built by Allied Chemical Corporation and others, at Barnwell, South Carolina, was scheduled to come into operation in the 1980s but it, too, encountered financial difficulties. They were seeking government financing to complete it when fuel reprocessing was eliminated from national energy planning as revised by the Carter administration.

It had been argued by scholars interested in reducing the dangers of nuclear proliferation that it would be safer if the plutonium produced in reactors was left in the original fuel pellets rather than reprocessed to create plutonium in a form ready for bomb making (Feiveson, 1976). It was also realized that reprocessing is no longer economically rewarding now that the government no longer buys the plutonium for its military program. The amount of enriched uranium yielded by reprocessing is only enough to increase the effective uranium supply by about 30 percent and it appears to cost no more to mine and enrich that much more uranium instead of chemically reprocessing it. In the new plan, the fuel cycle is thus replaced by a once-through policy.

Leaving the plutonium in fuel pellets changes the nature of the waste-disposal problem. Instead of handling high-level liquid wastes it will be necessary to provide facilities for storing and cooling the wastes in solid form, in large tanks of water similar to those at reactor sites.

It was desirable for the sake of world stability to eliminate the reprocessing step and it happened to be politically viable because the industrial attempt to perform reprocessing was foundering. Indeed, the genuine tie-in with world stability provided at the same time a plausible justification for the government to rescue the nuclear enterprise again by expanding its capacity to perform the fuel supply function in another form, uranium enrichment.

Even before the change of policy increased the demand for enrichment this was a troubled area and industry has declined to take over the burden. The physical separation of isotopes of almost equal mass, U-235 and U-238, in the preparation of enriched uranium fuel in the first place is a much more demanding process than the chemical operations of a fuel reprocessing plant. From the start the industry has been dependent on the separation services of the government's three gaseous diffusion plants, the original wartime one at Oak Ridge, Tennessee, and those at Portsmouth, Ohio, and Paducah, Kentucky, built in the rapid expansion of nuclear weaponry in the early sixties. The government has been trying to turn this part of the business over to industry but without much success, and the growing number of power plants is taxing the capacity of these government diffusion plants. Even before the policy change there was a program to increase their capacity by about 50 percent by 1981, at taxpayer expense of course, but this would not have been enough; more plants were needed. As Robert E. Fry, assistant administrator for ERDA, said in 1976, "If nuclear power is to grow in this country, we must build new enrichment plants. We must have many of them and fast. The next enrichment plant must be on line by 1983."

When reactor technology was released to American industry and to the world in the mid-fifties, there was concern that this would facilitate the proliferation of nuclear weapon-making capability. As one key precaution the details of the vital isotope separation process, specifically, the nature of the diffusion barriers, were kept a carefully guarded secret, and the diffusion plants were all government owned. Now, in the mid-seventies, this precaution has been abandoned and the diffusion-plant secrets have been divulged in the hope of inducing industry to get into the isotope separation business.

Recently in Western Europe, a rival process of isotope separation has been nearing final development. Called the centrifuge method, it functions on the same principle as a cream separator. It requires huge plants for an appreciable output, but this can be achieved by expanding small plants, which is impossible in the case of the diffusion method. One drawback of the diffusion method is that it consumes a great deal of power, about 5 percent as much power as is later produced in power plants from the enriched fuel it supplies. The centrifuge method is expected to use only about one-tenth as much power.

The advent of this new separation method means that divulging the power-plant secrets has not thrown away as much protection against nuclear weapons spread as it otherwise would have. The cat is partly out of the bag anyway.

However, this inducement was not enough to bring industry into the business. Big companies like General Electric and Westinghouse could not see their way clear to making such an enormous investment. To overcome this reluctance, in 1976 the administration proposed to bring the taxpayers' money again to the rescue, this time to the enormous extent of about a hundred billion dollars.

President Ford's proposal in the Nuclear Assurance Act was that government should guarantee investment in separation plants so that, if a venture is successful with either the diffusion method or the as-yet-unproved centrifuge method, the investor stands to win, but if not the government loses. Industry showed interest on these terms, but the act was not passed by Congress.

These difficulties are indications of the extent to which the nuclear power industry still needs government subsidy (BW, 1975; Welch, 1977). Some industrial leaders fully recognize this. Richard A. McCormack, a high official in General Atomic and formerly in Combustion Engineering, said in mid-1975, "The fundamental cause of our malaise is the nature of the nuclear business itself. . . . It was born in the government and consciously weaned by statutory and administrative policies to grow through an adolescence of government support. But the business was never firmly founded, and the magnitude of the problems of getting it established in the private sector was never fully appreciated." Emphasizing how technologically complex and capital-intensive it is and how it still needs government support he added, "Frankly, we are a sick industry."

8

Nuclear Power from Fusion

The sun is like a great furnace, drawing upon the potential energy inherent in the separation of its own component particles and converting this potential into heat energy as the particles draw closer together. The sun employs not only the long-range gravitational force that, even at astronomical, distances, pulls all particles of matter together, but also the short-range nuclear force that, at submicroscopic distances, pulls all nucleons closer together when they are in the process of forming atomic nuclei. In the case of gravitational attraction the conversion to heat is a simple concept: as the atoms and molecules are pulled toward each other, they move faster, ultimately resulting in the molecular or atomic agitation we know as heat. In the case of nuclear attraction of nucleons, the conversion is less direct. Their cohesion to form nuclei creates internal energy in the nuclei formed and permits the emission of radiations—alpha, beta, and gamma rays—which accelerate the atoms they encounter and thus raise their temperature. All this is happening in the complicated interior of the sun.

Fortunately, there is also a repulsive force between some nucleons; otherwise the sun would explode. There are two kinds of nucleons, neutrons and protons. The protons carry a positive electric charge. Two positive charges repel each other with a long-range force. The electric repulsion is weaker than the attractive nuclear force when two protons are very close together, closer than 10^{-14} meters or 10^{-12} inches, but the electric repulsion is stronger than the nuclear attraction at greater distances, where the nuclear attraction practically vanishes. This long-range electric repulsion prevents nuclei made up of a few nucleons from hitting each other unless they are thrown together very violently. Only if they do hit each other can the short-range attraction come into play and produce heat. This cardinal fact greatly slows the rate at which the nuclear attraction is converted into heat. It makes the conversion slow and steady, so that the sun does not explode but

138

instead, through its radiation of sunlight, provides us with a steady source of solar energy.

The combination of two very light nuclei, each consisting of two or three nucleons, into a slightly heavier nucleus with four or five or six nucleons, it known as fusion of the lighter nuclei into the heavier one. It can happen when the two nuclei are thrown together in an accelerator in a laboratory. It can happen when the nuclei are part of a very hot mass of particles in which all the particles move very fast; those which collide with especially high speed may undergo fusion. This type of fusion is known as a thermonuclear reaction. It happens at attainable temperatures between only relatively light nuclei like the nuclei of hydrogen, helium, and lithium, because heavier nuclei contain more protons and thus repel each other with even more force.

In an earlier book by the author (Inglis, 1973) much of this and what follows is explained in greater detail. It is emphasized that there are only two prospective sources of nuclear energy, fission and fusion, as shown in figure 41. The public should not expect science to pull more miracles of that sort out of the hat and should not use this hope as an excuse for squandering our energy resources. Figure 41 shows the energy per nucleon of all the nuclei

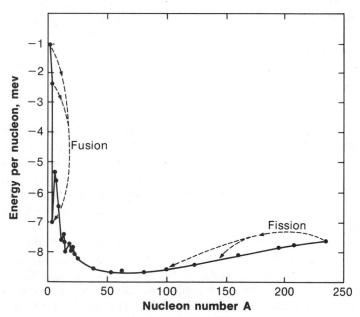

Fig. 41. Fusion and fission both release energy by transferring from a high end toward the low middle of the energy-per-nucleon curve. (Inglis, 1973.)

that the middle of the curve is lower than either end. If a heavy nucleus like plutonium with $A = 239$ splits into two medium-weight nuclei near the middle of the curve, the energy within the nuclei is reduced because the energy per nucleon is less according to the curve and there are the same total number of nucleons. This is one way of stating that the fission process releases energy. It does so because the two fragments push each other violently apart with electric repulsion. That is essentially why the middle of the curve is lower than the right end.

The left end of the curve slopes the opposite way because there are so few nucleons that the electric force is less important. Each nucleon is subject to the short-range attraction of more other nucleons as the number of nucleons A is increased. Thus fusion reduces potential energy (or makes it more negative on the scale of the graph) and releases kinetic energy that ultimately becomes heat because it brings more nucleons within the range of each other's nuclear pull. (In contrast to a heavy nucleus, a light nucleus is small enough that all nucleons in it are within the reach of one another's short-range attraction.)

If we consider nuclear forces alone and ignore the role of gravity, we might think that the hotter the interior of the sun is, the more thermonuclear reactions there would be, making the sun still hotter and heading for explosion. The sun's controlled and steady energy output comes from the fact that the hotter the interior, the more the sun expands against its gravitational pull, reducing both the temperature and the density inside and thus limiting the number of thermonuclear reactions.

The Containment Problem

Understanding the mechanism of this fascinating balancing act that Nature has devised to operate a controlled thermonuclear reactor, on a scale so astronomically huge that gravity comes into play, may help us appreciate the enormous diffculty faced by scientists who are trying to achieve a controlled thermonuclear reaction on a mere laboratory scale (and eventually on a commercial scale), without the help of gravity.

In the sun, the weight of the layers nearer the surface presses the reacting nuclei near the center together so forcefully that they have a high density in spite of their extreme heat. This causes very frequent collisions. On earth, two substitutes for gravity are being investigated as ways of holding the nuclei close enough together when very hot. These substitutes are magnetic fields and inertia. Neither strong steel tanks nor glass or ceramic bottles will do; they would vaporize instantly at the temperatures required or, if kept cool, would cool off the fluid they contain.

The nuclei to be contained are electrically charged and can thus be deflected by a magnetic field. A very hot collection of atoms and their nuclei at a low density and pressure, usually much less than atmospheric pressure, is known as plasma. A magnetic field can be arranged to surround a plasma in such a way that when the nuclei enter the surrounding field they are deflected inward. The plasma may be said to be contained in a magnetic bottle. Achieving a strong enough magnetic bottle that does not leak is one of the aims of thermonuclear research. Leaks have been evading the persevering efforts of scientists, working with international collaboration, for more than two decades. The leaks come mainly from instabilities that occur wherever the pressure of the plasma pushes on the magnetic field. Weird ingeniously shaped magnetic bottles have been invented in attempts to avoid leaks.

Inertia is the other substitute for gravity to maintain a high pressure, but it can do so for only a very short time. When a tennis ball bounces against a wall, for example, the wall pushes hard on the ball for a short time to overcome the ball's inertia, or, in other words, to change its momentum and there is fairly high pressure in the area of contact during that short time. In an implosion-type atom bomb, fission takes place in a central region of high density caused by high pressure based on the same principle: the material in the outer layers is propelled inward by a surrounding chemical explosion and the pressure at the center is high for a very brief instant while the material bounces back out, propelled by the energy released in fission. The hydrogen bomb involves an uncontrolled thermonuclear reaction in a very hot region of high pressure that is contained for an instant by the same inertial principle.

Now a new attack on the challenge of releasing thermonuclear energy on a laboratory scale is being made in a process called laser fusion. It can be likened to a sequence of miniature hydrogen bombs, depending on inertia for instantaneous containment. In one form, a solid pellet of the appropriate light atoms, largely lithium and hydrogen, is dropped and as it passes a certain spot, it is compressed by the pressure of an inward blast of laser light bombarding it from all directions. This creates a small implosion during which a thermonuclear reaction can proceed for the instant when pressure and temperature are very high. A burst of energy and neutrons is released that must be contained and employed usefully by the surrounding apparatus. The process is repeated at frequent intervals. One of the many hurdles to be overcome involves deriving more energy out of each burst than is required to make the blast of laser light, just as, in the magnetic method, it will be necessary to extract more than enough energy to maintain the magnetic bottle. It is difficult to find an adequate substitute for the sun's gravity here on earth.

The ultimate hope for fusion is to produce power using the deuterium of the sea as the main fuel. The common form of hydrogen has nucleon number $A=1$, a single proton as its nucleus. Deuterium with $A=2$ is the rare isotope, about one part in twenty-thousand of the hydrogen in seawater, but the seas are large and the supply practically inexhaustible. In laser fusion the $A=6$ form of lithium, which is the less abundant isotope comprising about one-sixth of ordinary lithium metal, would also be a fuel. Lithium is also an abundant element. Lithium may also be used with magnetic-bottle fusion as a surrounding blanket to absorb neutrons and produce extra power. Among the many questions to be faced is whether, if power could be produced at all, there would be more than enough not only to provide for the magnetic fields or laser light, but also to prepare the deuterium and lithium fuel consumed.

One advantage of fusion power over fission power is the potentially unlimited fuel supply. Another is that there are no fission products to be perpetually sequestered as high-level waste. There will still be problems with radioactivity produced by fusion reactors, but hopefully they will be less severe. Great quantities of radioactive tritium will be produced, and it is hard to contain this completely. Radioactivity will be induced in the materials comprising the reactor and in the lithium blanket, some of which will inevitably escape into the environment as is the case in the present fission fuel cycle.

The problem is so far in the future that one cannot assess with confidence the magnitude of these hazards, particularly when one cannot even predict the forms of the future reactors. At a correspondingly early stage in the development of fission reactors some of the present difficulties were not expected. We can, however, foresee that with fusion there will not be the great burden of radioactivity in a working reactor, unlike the accumulation in a fission reactor of fission products that pose a great hazard if a cataclysmic accident should release them to the atmosphere.

If it could be achieved without introducing undue danger, thermonuclear power would be a great boon to mankind. But this chicken should not be counted before it is hatched. Besides the technical problems, the economic uncertainties are so great that it would seem a miracle if it really comes to pass.

Yet there are workers in the field who are quite optimistic about ultimate success, some of them putting the odds as high as fifty-fifty. Those workers in the field know the most about the difficulties and the long history of trying to overcome them; perhaps we should all share their optimism. Yet we must be wary of the mutual-stimulation effect. Not only does mutual-stimulation between scientific workers foster rapid development of ideas, which is very beneficial, but it can also bolster optimism among a group of workers dedicating their lives to a project. Without such optimism perhaps the search could not go on, though the technical challenges, regardless of practical results, are exciting enough to inspire enthusiasm. As an example of optimism within the thermonuclear project in the past, in 1956 scientists (including this author) were being encouraged to join the project by exciting predictions that a demonstration of technical feasibility of controlled fusion was expected within three years. Now, more than twenty years later, that demonstration still seems close but elusive. In the meantime bursts of neutrons have been observed, proving that some fusion is taking place within the apparatus, but we seem a long way from getting as much energy out as is put in.

Once that stage of getting out more energy than is put in to produce it is reached, the engineering problems in the transition from laboratory miracle to commercial practicability will be enormous. How much more complicated the process will be than in the case of fission reactors is suggested by the length of time required to develop each one to the point of demonstrating feasibility. For fission, it was four years from the discovery of fission to the demonstrated chain reaction. For fusion, it will have been more than thirty years of intensive effort. Even if a commercial fusion reactor is engineered there would be, in addition to the economics, a question of the capability of commercial management in so complex a field. The performance of commercial manufacture and management in the much simpler fission field is not very encouraging.

All in all, the prospect for fusion must be judged very exciting but elusive. It would be folly to depend on it in energy planning.

9

Dollar Costs of Wind Power and Nuclear Power Compared

The decision whether to invest more resources in wind power or in nuclear power as a future energy source will depend on their respective anticipated financial costs as well as environmental and social costs and benefits. Cost projection for each of them is so unreliable that we cannot have great confidence in the estimates or be sure which will provide cheaper electric energy in the long run. It may be sensible to place more emphasis on the environmental and social criteria in making the choice. The installations should eventually pay off through the sale of power. Nevertheless, money talks in such decisions and it is important to consider the available estimates of those future dollar costs, despite their uncertainty.

Costs of Large Wind Dynamos

The cost of building a large wind dynamo is expected to be much less if it is mass produced under conditions of commercial competition to sell power than if produced singly for demonstration and testing. One must not be misled by the high price tags on ERDA contracts for demonstration machines, for these must also reflect the additional cost of making a successful bid and an incentive profit to be derived from the single machine. The most plausible estimates of quantity production costs should be those based on experience in building individual machines from which one may extrapolate from the actual costs to those of multiple production. Such estimates have been made in past years, some of them just after construction and testing projects in various countries, and allowance must of course be made for inflation to express the estimates in present dollars.

The principal available estimates of this sort, based on 100-kilowatt-scale or megawatt-scale wind dynamos of various designs,

TABLE 2. Cost Estimates of Large Wind Dynamos

Machine Rated Wind Velocity	Source of Estimate	Production, Number of Machines	Date of Estimate	Cost Estimate ($/kw installed)	Inflation Factor	1977 Cost Estimate ($/kw installed)	1977 Cost Estimate with Capacity Factor ($/kw average)	
							at 33%	at 40%
1.5 Mw based on Smith-Putnam 1.25 Mw (fig. 5) 30 mph	Putnam, 1945	6	1945	190	3.6	690	2,070	1,730
Postulated model with future improvements (fig. 6)	Putnam, 1945	Many	1945	100	3.6	360	1,080	900
6.5 Mw Twin Turbine (fig. 7)	Thomas, 1949	1,000	1948	68	3.2	220	660	550
3.65 Mw Rotatable Tripod (fig. 12) 35 mph	Golding, 1951	100	1951	115	2.8	320	960	800
200 kw based on Gedser, Denmark (fig. 8) 15 m/sec = 35.6 mph (Observed average, 6 m/sec)	Westh, 1975	100	1974	584	1.29	750	2,250	1,870
	Rosen, Daebler, and Hall, 1975[a]	100	1975	400	1.14	450	1,350	1,130
1 Mw Leaning Tower of Stuttgart (fig. 24) 10 m/sec = 22.4 mph (Assumed average, 8 m/sec)	Hütter, 1974	500	1974	320	1.29	410	1,230	1,030
6 Mw floating offshore, three-turbine (fig. 30)	Heronemus, 1972	Many	1972	155	1.65	260	780	650
6 Mw seabed based offshore	Heronemus, 1972	Many	1972	120	1.65	200	600	500
General Electric 1.5 Mw	JBF, 1977a	100	1975	432	1.14	492	1,475	1,230
3 Mw Schachle Tripod (fig. 57) 40 mph	Inglis, 1978 (appendix 4)	100	1978	205	.95	195	585	490

[a]Quoted by Sørensen, 1975.

are listed in table 2. Most of them are from some years ago but the last two are recent. The actual overall cost of the Smith-Putnam machine at Grandpa's Knob, including all development and engineering design, was $1,000 per installed kilowatt. The S. Morgan Smith Company engineers, shortly after termination of the Grandpa's Knob testing program, estimated that erection of six wind dynamos of essentially the same design, but slightly larger (1.5 megawatts), would cost about $190 per installed kilowatt. This estimate included the expense of access roads up mountainsides, which would not be encountered in Great Plains siting. Putnam and his associates also made preliminary designs of a simplified and improved production model that they estimated could be mass produced at about $100 per installed kilowatt.

This figure and similar estimates of the costs of other machines listed in table 2 are translated there into 1977 dollars by using the Wholesale Price Index (BEA, 1976) as an indication of the rate of inflation. Since the wind does not blow hard all the time, the capital cost that is more relevant than cost per installed kilowatt in indicating the cost of generating wind power is the capital cost per kilowatt of generated power averaged over a long time, that is, the cost per average kilowatt of wind-generated power.[1] This average power may be expressed as the maximum or rated power installed multiplied by a capacity factor, sometimes also called a plant factor. For wind dynamos in typical wind climates the capacity factor may be around 35 percent, meaning that as much energy is generated as if the machine were operating at full power a little more than one-third of the time and otherwise idle. For the particularly windy regions that are of most interest for exploitation of wind power on a grand scale, the western plains or at sea offshore, capacity factors of 40 percent or even 45 percent may reasonably be expected.

In the last two columns of table 2 the cost per average kilowatt is listed for capacity factors of 33 percent, applicable to widely distributed good sites, and 40 percent, appropriate for special regions of strong and relatively steady winds.

The rated power of a wind dynamo is not necessarily a good

---

1. For the sake of comparison with other units of power encountered elsewhere, one average kilowatt of capacity is equivalent to 8,766 kilowatt-hours per year or 10^7 British thermal units per year or 3.16×10^{10} joules per year or 5.2 barrels of oil equivalent per year. Then one quad (Q) per year or 10^{15} Btu per year is $3.34 \ 10^7$ average kilowatts or 33,400 average megawatts. One quad is 1.72×10^8 barrels of oil equivalent so one barrel (42 gallons) of oil or about one-fifth of a ton of coal is equivalent to 1,700 kilowatt-hours or 194 watt-years.

TABLE 3. Comparison of Some Wind-Dynamo Power Ratings and Sizes

Machine	Rated Power (kw)	Average Generated Power / Rated Power	Rated Wind Speed[a] (m/sec)	Average Wind Speed (m/sec)	Blade Circle Diameter (meters)	Blade Circle Area (meters²)	Ratio Rated kw / (meters²)	Ratio Average kw / (meters²)
Smith-Putnam	1250		13.4		53.4	2,230	0.56	
Gedser	200	26%	15	6	24	450	0.44	0.115
Stuttgart 1959	100	42%	8	4.6	34	908	0.11	(0.047)
Leaning Tower	1,000		10		80	5,000	0.20	

[a] meter/second (m/sec) = 2.24 miles/hour

indication of its size or cost. Rated power means merely the generating capacity of the electric generator that is driven by the rotating blades, which is determined not only by the average wind speed at the site but also by the extent to which the designer thinks it will pay to be prepared to exploit the occasional winds that are much stronger than average. This is illustrated by the comparison of wind dynamos in table 3. The Gedser machine is somewhat overrated, its rated wind speed being two and a half times the average wind speed. Indeed, there are seven other wind dynamos at less windy spots in Denmark of the same overall size as the one at Gedser, but with generators (for direct current) less than half as powerful, rated at 75 rather than 200 kilowatts. For a year of operation at Gedser the capacity factor, that is, the ratio of the average generated power divided by the rated power, was only 26 percent. This figure is so low merely because the turbine was generously equipped with an oversized generator that was seldom used to its full capacity. The figure would be higher for a machine more reasonably rated for a wind speed not so much greater than the average wind speed.

By way of contrast with this example, the two Stuttgart wind dynamos listed in the lower part of table 3 are designed and rated for the relatively gentle winds of the Western European interior, the rated wind speed for the larger one being somewhat greater in keeping with the slightly stronger winds higher above ground.

Both of the Stuttgart machines take advantage of modern engineering developments, with tapered and twisted composite fiberglass and plastic blades of good aerodynamic design. They, and some others listed in table 2, are thus more efficient than the Smith-Putnam and Gedser machines that are relatively crude in this respect. It is to be expected that future machines reaching the stage of quantity production will be among the best and most economical and their probable cost should not be judged by costs estimated on the basis of those earlier, cruder machines.

Reasonable estimates of the cost of future substantial production of large wind dynamos may perhaps best be made by critically contemplating the wide spread of estimates in each of the last two columns of table 2. One such estimate may be made from the 33 percent column for the cost in the widespread local regions of utility companies or industries, and another from the 40 percent column for very large arrays in especially windy regions.

Looking first at the last column based on a 40 percent capacity factor for these windy regions, we see that the three low figures, from $500 to $650 per average kilowatt, are for production runs of a thousand or more machines rated for especially strong winds, either aloft or offshore. While these may turn out to be the best estimates for high-wind regions, they are so much lower than other estimates for other wind regions that it is perhaps prudent to consider them slightly optimistic. They are made by capable engineers who had not personally had the experience of building similar machines. The high values, those above $1,100, are for smaller production runs of the two relatively primitive machines; the very high value, $1,870, having been revised down to $1,130. We may reasonably ignore these. The highest figure for a more modern design is $1,030 per average kilowatt for the German leaning tower design of Hütter at Stuttgart (fig. 24), based on experience with his 100-kilowatt wind dynamo in the nearby Swabian hills (fig. 10). It operated for about nine years, from 1959 to 1968, and achieved a capacity factor of 42 percent, being conservatively rated for a light-wind region. The leaning tower machine was designed to exploit the same wind regime at a higher altitude and should achieve the same capacity factor (a bit more than 40 percent) which would reduce the figure of $1,030 in the last column to $980 per average kilowatt. This machine, which was carefully over-designed, provides a good estimate of costs for a light-wind regime, but not for high-wind regions where wind power should be considerably cheaper. The last two lines of table 2 are late additions to the list and are discussed separately.

Of the remaining figures, whose average is $720, those below the average are figured for production by the thousands, but allowance should be made for their being perhaps a bit optimistic. Those above the average are probably high partly because they were figured for smaller production runs. A reasonable estimate based on all this experience is $750 per average kilowatt as the expected cost of erecting thousands of megawatt-scale wind dynamos in selected windy regions.

The last line of table 2 is the most directly indicative of the cost in 1977 dollars that should be expected as a result of industrial initiative. This estimate is based on experience with the construction of what in mid-1977 was the country's most powerful operating wind dynamo at Moses Lake, Washington (see fig. 57). The price quoted is for production of one hundred machines and might be even less in greater quantity. The contract price of the first one, just over a million dollars, includes $205 per kilowatt, delivered and installed, plus reasonable additional development and testing expenses for a first machine. The contract guaranteed generating 6000 megawatt-hours per year, corresponding to a capacity factor of only 23 percent, but with expectation of at least twice that. If the cost and performance turn out to be anywhere near this expectation, as seems assured, it will show that the figure of $750 per average kilowatt in strong wind regions, adopted here as a basis for discussion, is considerably too high and it will make the comparison with nuclear costs even more favorable.

The next-to-last line in table 2 gives a General Electric estimate that is much higher than this adopted figure. Yet it is the smallest of four such estimates made by government contractors, the other three being made by aerospace companies (JBF, 1977a). The pricing practices of these companies are probably influenced by the nature of aerospace and military government contracting. Their estimates thus probably do not reliably indicate costs to be expected in large-scale competitive industrial production of civilian hardware like wind dynamos. Unfortunately, being made at government expense, these estimates tend to be believed by government policy makers and are probably part of the reason they see wind power as not becoming very important until the next century.

A similar examination of the 33 percent column, giving more weight to the leaning tower figure and less to the low figures applicable to stronger winds, leads to an estimate of about $1,100 per average kilowatt for hundreds of machines in utility territories. This is not inconsistent with a statement attributed (Hill, 1976) to Louis Divone, director of the ERDA-DOE Wind Energy Conversion Branch, which says that it should be possible to get the cost down to $400 per installed kilowatt. This would correspond to about $1,200 per average kilowatt in the installations in user territories in which DOE has shown interest.

In short, we may expect megawatt-scale wind dynamos to cost about $750 per average kilowatt (1977 dollars) in huge arrays in special windy regions of the western Great Plains or offshore, and about $1,100 per average kilowatt when used locally over most of the country.

Transmission Line Costs

If the windiest parts of the country are to be exploited to serve densely populated areas, wind power will involve longer transmission lines than do fuel-consuming power plants. Although these plants are mostly near cities, some of their power output is transmitted several hundred miles between cities in regional grids as the local supplies and demands vary. Legislation has been considered that might establish a nationwide grid for this purpose which would considerably reduce the extra transmission costs for large-scale wind power.

Finding acceptable sites for nuclear power plants places them about thirty to one hundred miles from the cities they mainly serve. In the case of off-shore wind power off the eastern seaboard the distances may not be much more but from a huge wind-power system on the western Great Plains a typical distance may be five hundred miles. The most expensive part of such a power line is that near and in a city, where right-of-way is costly. This part would cost the same for wind or nuclear power. In comparing costs of wind and nuclear power, we neglect it for both.

The cost of high lines over open country is about twenty cents per kilowatt mile (Rockefeller, 1974). The power is generated and transformed to high voltage as alternating current and is transmitted as alternating current over moderate distances. Over distances of several hundred miles there are phase-stability problems, particularly with somewhat unsteady wind dynamos at one end of the long line, and direct current would be used. This is entirely practical and at these distances the transmission cost is about the same as for alternating current, the lower cost of the direct-current line compensating the cost of conversion from alternating to direct current and back (Forsythe, 1977).

There is a power loss of about 1 percent per hundred miles, so 1 percent of additional installed generating capacity is needed for each hundred miles of transmission. The cost of the wind dynamos, $750 per average kilowatt at a 40 percent capacity factor, corresponds to $300 per installed kilowatt. This is to be increased by $23 per installed kilowatt for each hundred miles of transmission, $20 for the transmission lines and $3 for the extra wind dynamos to make up the power loss. Again at 40 percent capacity factor, this is almost $60 per average kilowatt per hundred miles.

Including internal wiring of the wind dynamo array, the extra transmission for wind power from the western Great Plains, as compared with nuclear power, would require about six hundred miles to Chicago, or to hook up with some of the great hydroelectric installations of the West. At this distance the cost of wind power, in comparison with nuclear, is $1,100 per average kilowatt. With, say, four hundred miles typical of other big cities in the Midwest or for the Missouri River dams, it would come to about $1,000 per average kilowatt.

That figure is the cost without storage; the cost of power delivered to midwestern city grids from a supplementary source, subject to outages on days when the winds are light throughout a widespread wind-dynamo array. For delivery of dependable power on demand, about three-day storage should be included in the cost. This can be provided by partially underground pumped hydro-electric storage at an additional cost of about $450 per average delivered kilowatt including the cost of extra generating capacity to compensate for storage inefficiency as is discussed later in this chapter. This is reckoned on the assumption that the storage is at the consumer end of the long transmission line, near St. Louis, for example, where the cost of the wind dynamos and the high line capable of transmitting their varying power output is taken to be $1,000 per average delivered kilowatt. With storage at this end, the wind-power system can help meet the daily peak load without requiring that the transmission line be heavy enough to carry that load. If steady base power is desired instead, the cost can be somewhat less than this estimate if the storage is at the generator end of the line, so that the line need be only heavy enough to carry the average power, not the windy-day maximum.

To summarize, then, the cost of the installations to generate variable wind power in southern Wyoming or western Kansas and to deliver it to the great hydroelectric dams in Wyoming where no storage is needed, for example, is estimated to be about $1,000 per average kilowatt. The corresponding cost with storage for dependable wind power having a peaking capacity, delivered to the great midwestern city grids, comes to about $1,500 per average kilowatt, as we will see in the following discussion.

Need for Energy Storage

Power plants in general are undependable and some provision must be made to assure the continuity of municipal electric power supply over periods when a plant is underperforming. In systems of fuel-consuming power plants this is done by providing extra backup capacity in a large electric power grid, building more plants than otherwise needed to make it unlikely that too many will happen to be out of service at once.

Variation in load is a separate reason for providing excess generating capacity. The daily variation is the most serious one, the demand peaking in the early evening followed by a slack period late at night. The evening peak for three hours or so is typically 30 percent above the daily average. This rapid variation is superposed on a slow variation with the seasons, with periods of heavy demand in cold winter months and for air conditioning on hot summer days. While excess capacity is the usual method of providing for both of these variations, the power industry has shown in a few large installations that the peaking power for the rapid daily fluctuation can be provided more economically by means of energy storage than by extra power plants.

With wind power the problem is different. If we look to the future and consider a large electric grid deriving all its power from wind dynamos, the greatest reason for undependability is that the wind is fickle. While the system of many thousands of wind dynamos will be spread over an extensive territory, a period of gentle winds will affect many of them in one region at the same time. The most important periodicity of wind variation is a few days, typically a five-day weather cycle drifting eastward across the country. A period of two or three days of low winds may occur first in one region, then in another. If the system of wind dynamos were spread out over the country feeding into a nationwide grid, the total power would be quite steady. But if the dynamos were concentrated in one region such as western Kansas, most of them might experience gentle winds simultaneously for a couple of days. For this reason it would be useful to have an energy storage facility with a few days' capacity, rather than the few hours provided for peaking power in present systems. But once the storage capacity is provided for the longer-term wind variation, it can manage the peaking problem as well. It accomplishes two purposes at once, so the necessity for storage is not as great a disadvantage for wind power as it otherwise would be. It presents some disadvantage, for long-term storage is somewhat more expensive than short-term storage, but there is yet another compensating advantage; for some forms of storage it would not be too expensive to extend the storage to cover seasonal variations of supply and demand as well, as we shall see.

Between the high demand of midwinter and, in some parts of the country, the high air conditioning demand of midsummer, there are periods of lower demand. With fuel-consuming systems and only few-hour storage the amount of excess capacity needed is determined by the high seasonal demand, not by the average. With a wind-power system having several-month storage, the number of wind dynamos required would be determined by the yearly average demand. The storage capacity required would be determined to some extent by the seasonal fluctuations. Were it not for the air conditioning load in the summer, the storage capacity required would be much less than one might think because without it the curve of the seasonal variation in wind power available follows approximately the curve of varying seasonal demand. That is, the wind blows harder in winter when, because of the long nights and cold weather, there is the greatest demand for electricity. If, in the future, direct solar energy takes over the air conditioning load, as is appropriate since the sun shines warmest when that need is greatest, then there will be no need for additional storage capacity to accommodate the seasonal variations in the wind or in the demand (Andrews, 1976).

With a large wind-power system, then, the primary function of storage will be to provide needed supplementary power on days of less-than-average wind power after it has been stored up on days when the power generated exceeded demand. It is important to appreciate that under these circumstances most of the power goes directly from the wind dynamos to the consumer and only a rather small part of it goes through storage. To understand this, consider the simplified cases represented in figure 42. In case *a* the wind is assumed to blow hard enough to produce the maximum power 40 percent of the time and to be absolutely calm for 60 percent of the time, an unrealistic extreme of course. The average power is 40 percent of the maximum power, as indicated by the horizontal

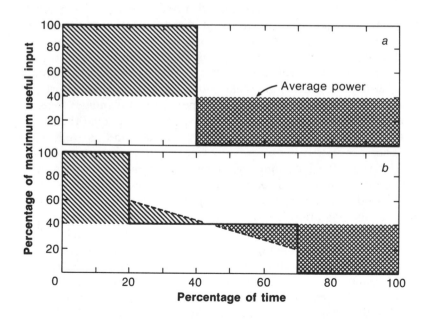

Fig. 42. Examples of the fraction of energy passing through storage from a variable supply such as wind power. The heavy line represents useful power generated. The shaded areas represent energy passing into or out of storage.

line, and the amount of energy that goes through storage, represented by the shaded areas, is 60 percent of the total energy. The upper left shaded area is transferred to the lower right shaded area through storage. Only the lower left clear area goes directly from dynamo to consumer when there is plenty of power for that and storage, too. (To complicate matters, storage is not 100 percent efficient and the energy plotted in the graph may be considered to be that part of the energy that is ultimately useful; or the energy generated reduced by a percentage that is wasted in the storage process. One hundred percent on the graph then represents somewhat less than the maximum power generated or rated power of the wind dynamos and the 40 percent average line is less than 40 percent of the rated power.)

In this simple example, constant consumption is assumed. If there were also short-term variations in consumption, the deviations of consumption from the average could be represented by a wiggly line above and below the horizontal average-power line. Transfers to or from storage would then be areas bounded by the wiggly line and would remain almost the same as with the straight line.

We might call figure 42a a one-step variation. Consider now the two-step variation of the solid line in figure 42b. This is a more realistic approximation of a real wind situation. The wind blows hard 20 percent of the time, blows with average strength 50 percent of the time and is nearly calm for 30 percent of the time. Here it is easily seen that 30 percent of the energy generated goes through storage. The horizontal part of the curve in the middle is unrealistic because it assumes the useful power generated to be exactly the average for part of the time, and the sloping broken line in figure 42b is a simple modification, still easily calculated., realistically assuming only that useful power is near the average for that part of the time. The area of each of the small triangles is 7.5 percent of the area under the average-power line and with this modification the shaded areas representing the energy transferred into and out of storage are each 37.5 percent of the total energy used. This plot including the sloping line is probably a close approximation to the curve of actual wind data which would also have a flat top at the upper left, as it attains the rated power, and an abrupt termination at the lower right end on reaching the cut-out speed below which no useful power is generated. It is a conservative assumption that as much as 40 percent of the power used may have to go through storage in a utility system dependent on wind power alone (Heronemus, 1972).

Costs of Energy Storage

Of the various possible ways for storing energy, the method that is best known and demonstrated in present commercial practice is pumped hydroelectric storage. The others that are under development are hydrogen storage with electrolysis and fuel cells, flywheels, compressed air, and improved storage batteries. While there is some hope that one of the others, in particular hydrogen storage, may be more economical and adaptable for use in wind power, the pumped storage method is with us now and its costs are known from present large-scale installations used to provide peakload power. In providing for the large daily and weekly fluctuations, it is expensive to build enough fuel-consuming power plants able to supply the maximum demand that usually comes in the early evening hours. The pumped storage scheme makes use of the generating capacity during slack periods (late night and early morning) to pump water into a high reservoir that can later be used as a hydroelectric source during the peak period. The same machinery works both ways, as a motor and pump on the way up and as a turbine and generator on the way down (fig. 43). Even though the overall efficiency of this process is only about 70 percent, so that 30 percent of the fuel is wasted compared with its use in a sufficient number of power plants, the capital costs for the storage system are less than they would be for additional power plants, which makes the method economically sound. It is especially valuable in connection with nuclear power plants because nuclear reactors deteriorate when frequently cycled between high and low power and are best used for base-load power only. It would be equally useful with wind dynamos to tide over windless days.

The largest storage system, at Ludington, Michigan, has a generating capacity of 1,700 megawatts, or almost that of two large nuclear plants, and the reservoir, at an average height of about 330 feet above lake level holds 27 billion gallons, enough to supply about nine hours of full-power discharge (Loving, 1971). It is built on a porous sand dune so its floor had to be paved with many layers of blacktop paving material. This is an unusual requirement since most of the plants have their reservoirs on solid rock. At Ludington, Lake Michigan serves as the lower reservoir. At Northfield, Massachusetts, it is the Connecticut River, which has the ecological drawback of causing sudden variations in river level. It is better to have a separate lower reservoir to keep the water in a closed system and avoid harming small fish and other marine organisms. The one at Northfield, for example, has its turbogenerators installed in a large

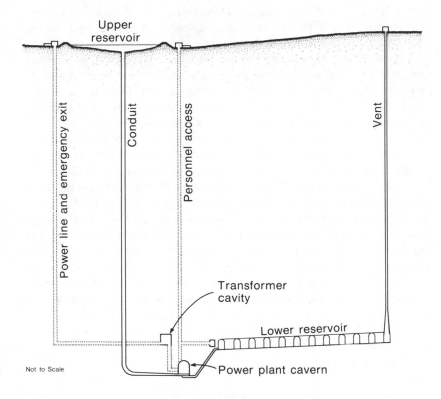

Fig. 43. Proposed New Jersey Power and Light Company underground pumped storage facility at Mt. Hope, about twenty-five miles northwest of Newark. (Not to scale.) Power 1,000 megawatts; head, 2,400 feet. (Adapted from *Energy*, 1977.)

powerhouse hewn deep within the granite rock of the mountain (fig. 27), demonstrating how nuclear power plants could also be built underground for added safety but at added expense. A proposal to build such a storage facility at Storm King on the Hudson River not far from New York City with its powerhouse above ground on the face of the mountain, has raised a storm of protest as an eyesore on a spot of natural beauty. So far, it is not being built.

During the peak-power application in a system of coal-fired and nuclear plants, nine-hour storage is enough. In a system of many wind dynamos in a limited region, three-day storage would be more appropriate for providing dependable power, smoothing out the fluctuations, and a month or more of storage capability would be better to accommodate seasonal fluctuations, though even one-day storage would be helpful, being capable of tiding over during a few days of less-than-average winds.

The cost of any storage system such as a pumped hydroelectric one may be divided into two parts. The conversion apparatus that converts the energy from electric energy to the form to be stored and back to electricity again, has a cost roughly proportional to the power input and output, measured in kilowatts. The other part, the storage capacity, in this case the energy stored in the high-level reservoir, is measured in kilowatt-hours. This part determines the length of time that the rated power could, if necessary, be delivered without replenishment. These components of the cost and the total construction cost for the four largest pumped storage facilities are listed in table 4 based on data from Newark (1976). The storage costs vary widely depending on the local geography. Building the upper reservoir on a large sand dune, as was done at Ludington, appears to be atypically expensive. Assuming that pumped storage would be used with wind power only in geographically favorable locations, we estimate costs more realistically by ignoring that first line of the table and considering only averages of the other three. This yields a total cost of $173 per kilowatt for a plant with storage capacity enough to supply the rated power for fifteen hours, according to the formula

$$\$137/kw + (15 \text{ hours}) \times (\$2.4/kwh) = \$173/kw.$$

Costs so derived for five different storage times are listed in the first line of table 5, where comparison is made with similar estimated costs of storage by other methods.

TABLE 4. **Cost of Large Pumped Hydroelectric Storage Plants, 1977 dollars**

Storage Plants	Capacity (Mw)	(hours)	Storage ($/kwh)	Power ($/kw)	Total ($/kw)
Ludington	1,675	9	10.9	161	260
Raccoon Mountain	1,370	24	1.61	112	150
Blenheim-Gilboa	1,030	11.6	4.45	129	180
Northfield	1,000	8.5	2.2	171	189

The number of favorable sites is fairly limited where the surface topography provides for two reservoirs sufficiently different in height and not too far apart. Two dams in a steep mountain valley are one possibility. The method can be made available more generally

TABLE 5. Estimated Costs of Storage Systems (in 1977 dollars per kilowatt)

| System | A | | | | | B Conversion Apparatus per Kilowatt | C Storage Capacity per Kilowatt-hour |
	3 Hour	9 Hour	3 Day	30 Day	3 Month		
Pumped hydro	144	159	310	1,865	5,320	137	2.4
Underground hydrogen		300	301	313	339	300	0.018
Underwater hydrogen		300	303	331	394	300	0.043
Liquid hydrogen		950	955	1,000	1,100	950	0.07
Flywheel	370	970	7,300	72,000	200,000	70	100

by use of underground reservoirs. In flat country, a surface lake or artificial pond may be used as the upper reservoir in conjuntion with a deep underground cavity, even an old mine, as the lower reservoir. Otherwise, both reservoirs may be underground, one much deeper than the other, with the powerhouse also underground (Scott, 1977). It is expected that this will not be more costly than present installations and has the greater advantage of not altering surface features. The Raccoon Mountain plant, the most economical of the four listed, has its lower reservoir underground.

Good rock for underground reservoirs is available throughout most of the United States except near the southern Atlantic and Gulf coasts where it is overlaid with too much alluvial material. Even with two surface reservoirs, it is advantageous to have the powerhouse underground (as at Northfield) and a little lower than the lower reservoir to assure a positive pressure and avoid cavitation at the turbines when they are used as pumps (Newark, 1976).

Hydrogen storage, which is not yet so well proved in practice, consists of using electric power, such as that generated by wind power, to separate pure water into hydrogen and oxygen by electrolysis, storing the hydrogen (and for some special uses perhaps also the oxygen) and then converting it back to electric power by means of fuel cells.

Hydrogen storage has the advantage that long-term storage is not very much more expensive than short-term storage, since much of the cost is in conversion from electrical power to storable hydrogen and back, but it has the disadvantage that the conversion process is relatively inefficient so that more power must be generated in the first place. Here we assume the efficiency to be about 50 percent, which is probably attainable but still is an industry goal. A further advantage is that pipeline transportation of hydrogen is cheaper than high-line transmission of electricity. This could be important in transmitting wind power from remote windy regions such as the western Great Plains to population and industrial centers. This system would also have an advantage in versatility of end use. The electrolyzers converting the wind-generated electricity to hydrogen would be near the wind dynamos and some of the hydrogen that reaches industrial centers could be used for direct domestic heating and industrial heat processes and some converted to electricity by fuel cells near the point of use. This system also solves the peak-power problem: the fuel-cell end of the conversion should be large enough to handle the peak demand for electric power, while the electrolyzer end needs to have the capacity to handle the fluctuations in the strength of the wind. The direct use of hydrogen for heating purposes requires some development but insofar as it can replace electric heating it will save the deplorable waste of converting fuel to electricity to heat.

While it is to be expected that costs of the critical components of a hydrogen storage system will decrease with future development and mass production, there are some rough estimates based on what may be considered present technology. Large-scale realization still contains some engineering challenges. For the price of converting alternating-current electric power (the kind usually generated) to hydrogen, a preliminary study by Heronemus (1972) cites a 400-megawatt electrolyzer designed for Oak Ridge National Laboratory by Allis-Chalmers; its 1970 cost being given as $35.5 million, or $81 per kilowatt of conversion capacity, roughly $100 in 1977 dollars. Most of this cost is for conversion from alternating current to direct current and could be avoided by designing the wind dynamos to generate direct current directly (as did the early Danish ones). (If seawater is used as input, there is an extra $5 per kilowatt for a distillation plant). As for converting the pure hydrogen back to electric power, he quotes a Pratt and Whitney Aircraft Division design of a fuel cell estimated to have cost $115 per kilowatt of capacity. This may be optimistic, for they have not been sold and no firm price tag has been put on them, although about sixty of them were leased experimentally to utility companies. Industry is reluctant to predict but it is probably safe to assume $200 per kilowatt in 1977 dollars for future mass production, pure hydrogen to alternating current. Thus one may estimate the cost of the apparatus for the round trip from electric power to hydrogen gas and back, at $300 per kilowatt in 1977 dollars, with considerable uncertainty attached to this figure.

It is envisaged that the hydrogen will be stored either as gas under high pressure or as a liquid at low temperature. One first thinks of storing high-pressure hydrogen gas in large steel tanks thick enough to contain the pressure, but this turns out to be very expensive. It is much more economical to use the high pressure provided by nature under thick layers of either rock or water. Underground storage of high-pressure natural gas is commonplace in industry, and sometimes the stored gas contains an appreciable fraction of hydrogen gas. Depleted oil wells provide suitable cavities; where they are not available cavities may be made in salt beds by dissolving out salt with circulating water.

The cost of underground hydrogen storage has been estimated by modifying the actual costs for natural gas storage in consideration of the different energy density of an equal volume and pressure of hydrogen. The estimate is $0.01 in 1970, or $0.018 in 1977 dollars per kilowatt-hour for storage in depleted oil wells or about three times that in dissolved-salt caverns, both at a pressure of 1,000 pounds per square inch (F.P.C., 1971). At this rate the storage component for even one-year storage would cost only about $65 per installed kilowatt, much less than the associated conversion apparatus at about $300 per kilowatt, making a total for the one-year storage system at about $350 per kilowatt. To smooth out annual fluctuations as well as the short-term variations, a three-month storage system should be ample and our estimated cost for this is about $310 per kilowatt. This would be enough to carry over six months or so of less-than-average winds since even if they produced only half the average power, for example, only the other half of the power consumed would be drawn from storage.

As a part of his proposal to exploit the strong offshore winds off the eastern seaboard, Heronemus has proposed storing hydrogen under the pressure of the deep sea just off the continental shelf, where the depth is about 7,000 feet or about 2,300 meters. This involves construction of large concrete hulls from which seawater can be displaced from beneath the pressurized hydrogen, the hulls with their ballast being heavy enough to keep them on the sea bottom despite the buoyancy of the gas. The cost of this storage, including the cost of an adequate compressor, is estimated at $0.026 per kilowatt-hour, in 1972 dollars (Heronemus, 1972), or about $370 per kilowatt-year, in 1977. The cost of a three-month storage system at this rate would then be about $300 for conversion apparatus plus about $90 for storage or about $390 per installed kilowatt. This cost is more than for storage underground in depleted wells but a bit less than in dissolved-salt caverns.

Liquid hydrogen storage is intrinsically more expensive and less efficient than pressurized hydrogen storage and will probably never be competitive in large-scale systems. The inefficiency comes from the necessity of a cryogenic apparatus to liquify hydrogen, which is expensive both in energy and in capital, in addition to the apparatus for the electricity-hydrogen-gas-electricity conversion that is needed for pressurized hydrogen storage. Compressing the gas is simpler than liquifying it. Whereas the cycle electricity-pressurized gas-electricity may be expected after some development to have an efficiency of around 50 percent, that for the cycle electricity-liquid hydrogen-electricity has been estimated at 29 percent. The cost of the liquification plant is estimated (Hallett, 1968) at $1.05 per 1,000 Btu per day or $45 per kilowatt of capacity in 1968 dollars, $86 per kilowatt in 1977 dollars. The storage element of a liquid-hydrogen storage system requires good thermal insulation to minimize evaporation loss rather than mechanical strength to withstand pressure. In spherical tanks with vacuum-perlite insulation the evaporation loss would be about 5 percent for three month storage and the cost $0.14 per kilowatt-hour of capacity in 1971 dollars, $0.24 per kilowatt-hour in 1977 dollars (Penner, 1975).

There are possibilities for electrochemical storage of energy that include the ordinary storage battery. A more promising modern form known as the Redox flow cell involves ion exchange through a membrane between $TiCl_n$ and $FeCl_n$ electrolytes. Estimates counting on future advances place the cost somewhere between $200 and $330 per kilowatt for nine-hour storage capacity, in 1977 dollars. This includes alternating current to direct current conversion (Warshay, 1975).

Another storage medium that has been discussed but not yet put into large-scale use is the mechanical energy of a rapidly spinning flywheel. A well-designed flywheel in an evacuated chamber to avoid air friction can store energy for quite a long time. The limitation on the energy that can be stored per kilogram of flywheel is the strength of the material of which it is made, for it can be spun only so fast that there is no danger of its bursting from centrifugal force. New fibrous materials bound together in plastic are much better than steel in this respect and introduce the prospect of economical energy storage by this method (see fig. 28).

Flywheel storage will probably have the advantage of greater efficiency than pumped hydroelectric storage and surely greater than hydrogen storage. An estimate by Post and Post (1973) of efficiencies and costs probably attainable in the not-too-distant future suggests efficiency above 90 percent and a cost amounting to about $165 in 1977 dollars per kilowatt for three-hour storage. This estimated price is competitive for the short-term storage needed for daily peaking power, but the method becomes prohibitively expensive for long-term storage, as is indicated in table 5.

In that table both the short-term and the long-term storage costs for the various methods are compared, estimated mainly from the sources just discussed. Since most of them are based on anticipated future developments, there is necessarily considerable uncertainty in these estimates. A recent commercial assessment sponsored by ERDA (Newark, 1977) handles that uncertainty by stating them only within quite wide ranges. The figures for pumped hydroelectric storage in table 5 fall well within the range stated. However, that assessment indicates that flywheel storage might cost up to four times as much as the Post (1973) estimate, emphasizing the conclusion that this method is unsuitable for wind-power applications.

The figures for flywheel storage in the last line of table 5, represent a compromise between the two sources. That recent assessment discusses the hydrogen method only with above-ground storage pressurized in steel tanks, which is much too expensive, and suggests that the conversion apparatus might cost up to three times the figure 300 in table 5.

If these storage methods were available at approximately these costs, the choice between them would of course depend on the uses contemplated. As a simple example, let us consider an isolated electrical network supplied only by a large number of wind dynamos in a limited region, all simultaneously subject to about the same variation of the wind, and in a wind regime requiring 40 percent of delivered power to have passed through storage. We may speak of the capital cost of the system, storage and wind dynamos included, per average delivered kilowatt at some season of the year with a three-day storage capacity, enough to average over daily and weekly fluctuation of wind and demand; or we may speak of a year-round average delivered kilowatt with three-month storage, enough to smooth out seasonal fluctuations.

The estimated combined costs per suitably averaged delivered kilowatt of the wind dynamos and storage capability in such a system for various storage methods and storage times are displayed in table 6, for realistic assumed efficiencies of the storage cycle. There is a trade-off between storage cost and storage efficiency that affects the cost of wind dynamos because with inefficient storage it is necessary to build more wind dynamos to make up for the energy wasted. If the storage facility is near the distant city being served, the output of all the wind dynamos must be transmitted that distance. This makes the capital cost of the wind dynamos and power lines, according to our estimate, $1,000 per kilowatt generated and so transmitted as unsteady power to the city with its storage facility. The storage inefficiency wastes some of the power at that end, adding to the cost. In estimating the total capital cost, we have assumed a capacity factor of 40 percent for the wind dynamos, as in the last column of table 2.

Table 6 illustrates the trends for various storage efficiencies and storage times. It shows that pumped hydroelectric storage, a technology already in use, is best for storage capabilities of a week or less; that, when developed, hydrogen storage promises to be more economical for seasonal storage; and that flywheel storage will probably be competitive only for a few-hour storage requirement for daily peak shaving or gust smoothing or both.

TABLE 6. Estimated Cost of Wind-Electric System with 40 Percent of Consumed Power Going through Storage

System	D In-out Storage Efficiency Assumed (percent)	E Extra Generating Capacity (percent)	F Cost of Wind Dynamos and Transmission per Average kw Delivered at $1000/kw Generated	G Total Cost Including Storage per Average Kilowatt Delivered (1977 $/kw)				
				3 Hour	9 Hour	3 Day	30 Day	3 Month
Pumped hydro	70	14	1,140	1,284				6,400
Underground hydrogen	50	40	1,400		1,700	1,701	1,713	1,740
Underground hydrogen	40	60	1,600		1,900	1,901	1,913	1,940
Underwater hydrogen	50	40	1,400		1,700	1,703	1,730	1,800
Liquid hydrogen	29	98	1,980		2,930	2,935	2,980	3,100
Flywheel	90	4	1,040	1,400	2,000	8,300	73,000	200,000

N.B. $A = B + C(t/hr)$; $E = (100/D - 1)40$; $F = 1000(1 + E/100)$; $G = F + A$.

This illustrative example is extreme in several ways. Wind dynamos will presumably be used in systems having other sources of power with fluctuations of input not dependent on the wind, making dependence on storage less. The illustration shows that, in the future, wind power with reasonable storage could stand alone as a source of electric power, but here also the example is an extreme one because in this case wind power from several widely separated windy regions would be fed into a national electric power grid and weak winds would not often occur everywhere at once. Thus a much smaller fraction of the energy would have to go through storage.

The illustration of table 6 uses a simplified method of averaging that is sufficient to show the general trends of cost. With records of actual wind variations at a given spot and with realistic curves of energy consumption, it is of course possible to make a more elaborate computation of the fraction of energy going through storage. Such computations tend to show that for wind dynamos the size of the one at Grandpa's Knob, in a single favorable windy region, the fraction of the delivered energy going through storage is considerably less than the 40 percent assumed in table 6. The fraction is smaller for very high towers that take advantage of the steadier winds aloft and for the relatively small individual wind turbines having a lower cut-in speed in order to make use of gentler winds. For example, in suggesting arrays of very many relatively small, 20-kilowatt, wind dynamos and using wind data only at 100 feet of altitude from three specific localities, Heronemus (1975) gives these small values of the percentage going through storage: in Iowa, 3.7 percent, in Upper Michigan, 8.6 percent, and over Lake Ontario, 10.5 percent.

If wind power from a limited region is ever to be used as sole input of a dependable electric network making storage as important as implied in table 6, it might be economical to use large numbers of small turbines. But one would have to decide whether the presumably greater construction and maintenance costs per kilowatt would be offset by the savings in required generating and storage capacity. For the more immediate prospective uses of wind power combined with other sources and feeding into extensive power grids, storage is a less important factor and the larger, megawatt-scale machines are probably economically preferable.

Uses Not Requiring Continuous Supply

Municipal electric power and many industrial power uses require dependable electric power on demand, and for these applications the cost of power must usually include storage costs as we have just discussed. Even with the cost of storage, wind power is found to be economic by a sufficient margin to cover any uncertainties in the cost estimates. But there are other applications where interruptions in power supply can be tolerated and here the economy of wind power is still more impressive.

A rather large part of our electric power is used for space heating. For this it does not make sense to store energy in a form such as pumped hydroelectric storage to be reconverted to electric power and then to heat. It is cheaper and more efficient to store it as heat at the point of end use. With local wind dynamos for single dwelling or community use (fig. 25) the wind power when available heats up the storage medium that can then be used to supply continuous heat as needed. With large-scale wind power from remote windy regions feeding commercial grids, it should be possible with modern communications systems both to control the heating of storage facilities in individual dwellings and to adjust the rates charged for metered power according to the availability of wind power. Even without wind power, it is likely that such a system will be devised to reduce peaking demand.

A lot of power is used for pumping water and here simple storage of water can tide over windless periods. The American farm windmill stores water in watering troughs for the cattle and in a tank for household use. The traditional Dutch windmill, to keep the land from flooding, did not need to pump continuously. The same is true of windmills pumping water for irrigation (fig. 16). Here irrigated earth itself stores water over windless periods just as it does between rains on rain-watered farms. All these historic water-pumping windmills were located where the pumping was needed and pumped water directly without using electric transmission of power. In the new renaissance of wind power this will be true in some cases again and the mass production of large wind dynamos for other uses may be accompanied by production of water-pumping windmills using the same wind turbines but with pumps in place of generators and gears.

In the large-scale irrigation that is important in American agriculture today, it will probably be more convenient and economical to use wind power transmitted by electricity driving pumps where needed. For this, large wind dynamos of the same type used for other purposes can be used either with separate transmission lines from nearby windy locations or through general electric grids with special arrangements for timing the irrigation pumping when the wind is strong.

Some other large industrial uses of electric power do not need a dependably continuous supply and could be located near regions of favorable winds to make use of wind power without storage. In much the same way that the aluminum industry has been attracted to regions of cheap hydroelectric power in the past, such power intensive industries could, in the future, be located either where there is wind power alone or where wind power supplements some hydroelectric power.

One of the travesties of our precarious civilization is the extent to which agriculture is dependent on fast-disappearing fuel sources. Most fertilizer is now made from petroleum and natural gas but could be made from ammonia produced by wind-generated electricity without storage. About one-third of the power used to pump water for irrigation is supplied by natural gas. This is one of the most compelling reasons why we should have a federal project to get large windmills in mass production soon. For this the cost of wind power without storage is relevant, leaving it not only cheaper but much cheaper than fuel-consuming sources of power.

Costs of Nuclear Power

The costs of nuclear power plants have escalated so fast that it is not possible to extrapolate into the future and give a reliable estimate of what a new generation of nuclear power plants and their associated equipment will cost (Morgan, 1977). In the light of the upward trend there is little doubt that they will cost more, figured in constant dollars, at least in the near future. When the contracts without fixed price for the present generation of large power reactors were first being made in about 1965 the cost was quoted at about $100 but turned out to be about $300 per installed kilowatt. In the next four years those figures rose to over $200 and almost $500, respectively, per installed kilowatt (all in 1973 dollars), as shown in figure 44. That these expected costs of the lower curve continued to rise drastically is shown in figure 45. There comparison is made with the corresponding cost trends for coal plants and refineries to show that the steepness of the rise is characteristic of the especially demanding nuclear technology, not of less sophisticated large installations that would include windmills. A 1,000-megawatt nuclear power plant ordered in 1975 was expected to cost $868 per installed kilowatt, compared with $226 in 1969 (Rhodes, 1976). In 1977 this extrapolates to an expected cost, in current dollars, of either about $1,100 or $1,400 per installed kilowatt, depending on whether the rise is treated as linear or exponential. The 1977 expected price may be taken to be at least $1,100 per installed kilowatt.

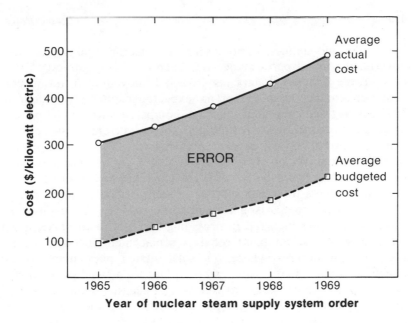

Fig. 44. Gross misestimation of reactor costs has troubled the nuclear industry. The bottom line represents estimated costs of a reactor ordered in a given year, and the top line the actual costs ordered or expected (1973 dollars). (Bupp, 1974.)

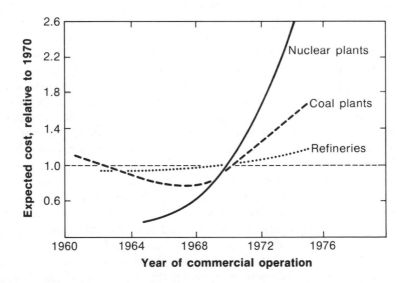

Fig. 45. The cost trend for nuclear plant construction has outdistanced coal plant costs and oil refinery costs, indicating that labor and other construction costs do not tell the whole story of nuclear plant cost increases. (Bupp, 1974.)

In the mid-sixties an anticipated plant factor of 80 percent was advertised. Actual performance has been averaging around 55 percent. Three hundred dollars per installed kilowatt with an 80 percent performance then meant an expected capital investment of $375 per average generated kilowatt, compared with $2,000 per average generated kilowatt resulting from the current $1,100 per installed kilowatt and 55 percent performance. This capital investment for the power plant itself, about $2,000 per average generated kilowatt, is the figure with which the expected capital costs of the generating plants for other sources of energy should be compared.

A separate comparison is then to be made of the capital and other dollar costs incurred in providing fuel and other services for the plants. A nuclear plant requires replacement fuel and waste-disposal expenditures whereas a solar-related plant does not. It might be argued that a solar steam plant or a wind-power array would require more laborious servicing because of the large size of the system and this might partially offset the fuel costs of the nuclear plant. It must be recalled, however, that the servicing of a nuclear plant is inordinately time-consuming and expensive when it involves welding or other demanding repairs in regions of high radiation intensity, requiring a relay of workers, each working a short time while receiving his permissible dosage. It is thus plausible that maintenance costs, which are in any case not a very large item, may be roughly the same in the two sources. We may assume that the sum of this and the other usual annual charges, including interest, insurance, taxes, and overhead, is about the same percentage of capital investment in the two cases so that a comparison of capital investment, after also allowing for fuel costs, is a fair comparison of the total dollar costs of power in the two approaches.

As capital costs rose, the cost of uranium oxide jumped by more than a factor of three, from about $8 to about $30 per ton, while the price of oil jumped by a similar factor, in about 1973. If uranium costs had risen and those of oil had not, this could have put nuclear power out of business. However, the two jumps were probably not unrelated because the two fuels are in competition and when one gets expensive and scarce, there is more demand for the other. There may also be more insidious interrelationship, since some of the big energy conglomerates are in both fields.

The future cost of uranium will depend both on how many nuclear plants are built and how much tax money and venture capital goes into stimulating exploration and development of new uranium sources, as well as unpredictable luck in discovering new deposits. Some experts expect that even the present reactors and those under construction will exhaust the high-grade uranium deposits before the end of the century, while others predict that new discoveries and the more expensive use of medium-grade and low-grade ores will make it possible to fuel present reactors and perhaps five hundred additional light water reactors that might be built by the end of the century (Bethe, 1975). Here, then, is another uncertainty in future costs of nuclear power but it is perhaps less important in the near future than the uncertainty of capital costs.

Nuclear power is notoriously capital intensive and accordingly fuel costs at present account for only about a quarter as much of the cost as do capital costs, or about a fifth of the total cost. This means that the power cost is rather insensitive to the first increases in fuel cost: a doubling of fuel cost, for example, would increase the cost of power by only about 20 percent.

Capital cost is reflected in the cost of power through the interest and profit return on investment that is charged against the capital investment. In comparing costs with another power source that has no fuel charge we may conveniently define for a nuclear plant an effective capital cost that is expanded to include, in effect, the fuel charge. That is, it is a fictitious capital cost great enough that, figuring in the usual percentages, one would get the actual cost of power without adding a fuel charge. (We combine operating costs with capital costs for both nuclear and wind power.) With the fuel accounting for a fifth of the total, this effective cost is about 1.25 times the actual capital cost. With the future capital cost of nuclear power plants estimated at $2,000 per average kilowatt, this makes the effective cost, for comparison with capital costs of wind power, about $2,500 per average kilowatt. In this figure, in 1977 dollars, the fuel is taken into account.

In view of the Carter administration's policy to try to avoid counting on breeder reactors for our future power supply, there is now less interest in a similar comparison between the costs of wind power and breeder-reactor power. Such a comparison has even greater uncertainties, particularly in view of the history of earlier gross underestimates of present costs of nuclear power. It appears to leave breeder reactors no more favorable than present reactors in comparison with wind power (Cochran, 1975). In 1975 the General Accounting Office, considering the larger breeders it assumed would follow the Clinch River 350-megawatt prototype, quoted a cost of a 1,000-megawatt breeder at about $1.5 billion, or, in 1977 dollars, about $1,700 per *installed* kilowatt, to be compared with the input figure we have used of $1,100 for reactors of the present type. This estimate would make it more expensive than those, even though it breeds and does not require financing of replacement fuel.

Comparative Costs in Large Municipal Systems

Neither a wind dynamo nor a nuclear power plant can be depended on to produce power at any given time. We have discussed the cost of storage to supplement an array of wind dynamos in order to answer the question asked in utility company advertisements, "What do you do when the wind stops blowing?" One might also ask, what do you do when the nuclear plant stops functioning? For the present generation of nuclear plants, the answer is that you start using more fossil fuel in the coal- or gas- or oil-fired plants that are part of the same electric power grid and act as backup for the nuclear plant. If occasionally some of them are out of commission at the same time, you may have a brownout. This is just as possible for wind power with the same fossil-fuel backup, without adding more than the cost of very short-term storage to smooth out the hour-to-hour gustiness of the wind.

For a promising initial large-scale use of wind power the fair comparison with nuclear power is without any storage for either, since large but still limited amounts of wind power can be used effectively in conjunction with present hydroelectric facilities or as supplementary power in existing power grids without paying for storage. In this comparison our present estimate makes nuclear power more than twice as costly as wind power, roughly $2,600 versus roughly $1,000 per average kilowatt. In more extensive future applications the two systems will depend on storage or reserve generating capacity in different ways. For such applications one should take
storage costs into account in comparing them.

If all goes well with a nuclear power plant it functions continuously for about a year and then has a six-week "outage" for replenishing fuel. In a system with several plants, these planned outages can be scheduled one at a time in periods of light load, in the spring and fall when there is neither heating nor air conditioning load. But experience shows that all does not go well as witnessed by their approximately 55 percent capacity factor. There are commonly unscheduled outages that may last for months, sometimes even years.

Fossil-fueled plants also have unscheduled outages, albeit somewhat less frequently. This means that in a system of many power plants, either an actual mixed plant or a hypothetical all-nuclear one, there must be extra generating capacity to carry on when one or more of the plants are out of order. In a mixed plant the nuclear plants are normally backed up by extra fossil-fuel capacity, having lower capital costs but higher costs for fuel when needed. Because it may happen that several of the plants in a system may be out at once, more generating capacity is needed than enough to supply the average power, enough that a brownout because of simultaneous outages will seldom happen. This is a statistical matter; one can never be quite sure that all or almost all of the dozen or so generating plants in a system will not be out at once. Here, large-scale wind power with its thousands of smaller and simpler generating units would have a great advantage. Big deviations from an average are much less likely with large numbers.

The shortfall from an unscheduled outage in one nuclear plant may sometimes be mitigated by postponing scheduled fuel-changing outages in other plants. However, this does not appreciably reduce the need for extra capacity since this must be sufficient to back up unscheduled outages during the peak demand period in midwinter when the fuel-changing outages would not normally be scheduled.

Wind power also is subject to a variability measured in months due to the seasonal variability of the wind. In most places the average wind power available in late summer reaches a low about half of the winter peak. Aside from the summer air conditioning demand, this very nicely matches the demand for electric power because of both the winter heating demand and the long winter nights. In some southern parts of the country the summer air conditioning peak demand is higher than the winter peak but this will hopefully be increasingly met fairly soon by solar cooling. Without the air conditioning summer hump, it could be claimed that the variation of wind power is much better correlated with the demand curve than is the uncorrelated occurrence of nuclear-power outages. Until solar cooling takes over, however, we may reasonably consider these several month-long variations as about equal disadvantages of wind and nuclear power, requiring similar amounts of back-up capacity that need not be counted in comparing the two.

Besides this, wind power has a short-term variation of supply that must be considered; the occurrence of a few hours or days of light winds or calm during the course of the weekly variation of the weather. If the wind-power component of a system is to deliver dependable power on demand, something like three-day storage is needed, enough to carry over about a week of less than average windiness. It would then also be able to help meet evening peak demand. A fair comparison of nuclear and wind power typical in a midwestern city would then require three-day storage as well as appropriate transmission lines for wind power and perhaps three-hour storage or an equivalent backup generating capacity to meet evening demand for nuclear.

Three-hour storage, in conjunction with wind power, would cost about $284 per average kilowatt according to table 6, and the figure would be a bit more, roughly $300, for three-hour storage or competitive backup capacity with nuclear power. This added to $2,600 gives $2,900 per average kilowatt for nuclear power with the capacity to meet evening peak demand, as compared with $1,450 per average kilowatt, from the same table, for wind power with three-day storage and 400-mile extra transmission. Under these conditions, wind power is estimated to be 50 percent as expensive as nuclear power. If we favor nuclear power by omitting the evening peaking consideration, the wind power cost is instead 55 percent of nuclear. This still leaves an ample margin to cover uncertainties in the input numbers and supports the conclusion that, even with each performing in an independent system, large-scale wind power would be appreciably more economical than newly installed nuclear power.

This is, however, an answer to a somewhat academic question, for neither source is expected to stand alone in a pure system. In more realistic mixed systems there will be considerably less dependence on storage, perhaps cheaper storage in the future, or even no dependence on storage if hydroelectric generating capacity is included in the system. Then the relevant cost comparison will be closer to that with no storage, leaving wind power much more economic than nuclear power.

Another type of comparison concerns a small contribution of either nuclear power or wind power to a large system otherwise dependent on fossil fuel, for example, one with just one nuclear plant or an equivalent generating capacity consisting of several hundred large wind dynamos close together in a single wind regime undergoing the same minute-to-minute wind fluctuations. The fossil steam plants cannot be expected to readjust output quickly enough to compensate for less-than-hourly changes in wind. To be equivalent to the nuclear plant, the wind array should have a little short-term storage, say about three-hour storage, at a cost of about $284 per average kilowatt, according to table 6. This still leaves an estimated cost of about $1,300 per average kilowatt, much less than for nuclear power. Sørensen (1976) has presented an interesting discussion of this aspect of the comparison, but with rather different cost estimates.

Quite another type of cost comparison between wind power and oil-fired capacity has been made by Coty and Vaughn of the Lockheed-California Company (Coty, 1977). They calculate under what conditions a utility company that has already made the investment in enough steam plants (some oil-fired) to meets its needs, would find it economic to install additional wind-power generating capacity simply because it saves oil whenever the wind blows. No storage is needed. It is purely a fuel saving proposition, balancing capital and servicing costs of wind power against just the cost of oil saved. They postulate a production run of a number of 2-megawatt wind dynamos with blade spans of 260 feet and express confidence that a company such as theirs could design and manufacture these now without further demonstration and testing programs. This makes our proposal that there should be government-sponsored program testing several models against one another at once before starting production runs look conservative.

Their wind dynamo is designed to take advantage of winds averaging fifteen miles per hour, less than at Grandpa's Knob, and the area per megawatt of the blade circle is 40 percent greater than that of the Smith-Putnam machine. This adds to the expense, but even so, their postulated cost for such a wind dynamo, $1.52 million or $760 per installed rated kilowatt in a production run of 100 units, is high compared with the costs we have discussed above. They use a typical observed wind-duration curve in calculating annual output. They study the variation of unit cost with change of the number of wind dynamos in the original production run, over which the design and set-up costs are spread and to which a learning curve is applicable. Since theirs is a comparison of capital costs with fuel costs, the annual costs such as taxes, depreciation, interest, and maintenance are included and these are different for various types of utilities. Their conclusion is that the operation would be economic with a production run of 260 or more units for a private company with no public incentives, or 30 units if there were guaranteed loans and freedom from taxes at least for this first run.

This comparison of wind power costs with oil fuel costs suggests an attractive way for large-scale wind power to get into production but is not directly related to a comparison of wind with nuclear power because nuclear power has considerably lower direct fuel costs, though higher capital costs, than oil-fired or even coal-fired steam power. This relatively low cost paid by the utility company for the fuel component of nuclear power is exploited in the way some utility companies have falsely advertised the claim that the use of nuclear power is saving customers money compared with using coal. In at least one case this claim was based solely on fuel costs saved with no consideration of capital costs when, with the latter included, nuclear power was estimated to be about 50 percent more expensive than coal-fired power (Comey, 1975d).

Another attractive use of wind power with storage in a large system is to supply peaking power. In this case it replaces more expensive standby generating capacity (Coste, 1977).

In any such discussion as this it must be appreciated that there are no dependable figures for future power costs. One can make careful but uncertain estimates, using general judgments and avoiding excessive details, to come up with an idea of which source will probably be the more economical. One could give an estimated range within which each figure is expected to lie, but the width of the range is another rough estimate. It may seem inconsistent to give a figure as apparently exact as $1,703, for example, in table 6, when it might lie anywhere between $1,500 and $1,900 or so, but the purpose of using four figures is to show that it is only a little larger than the figure to its left in the table. The relative numbers have more meaning than the absolute numbers. This is probably true also of the comparison between wind power and nuclear power; the estimates for them might be both low or both high because of changes in general construction costs.

Enthusiasts for nuclear power, expecting its price rise to level off, may claim that the estimated cost of nuclear plants ordered in 1977 should be 20 percent lower, say $900 rather than $1,100 per installed kilowatt with 60 percent plant factor, or $1,500 per average kilowatt without counting the cost of nuclear fuel, meaning $1,900 with the cost of fuel included as before. Skeptics about wind power may claim that the price of mass-produced wind dynamos should be about 25 percent higher than our estimate, say $950 rather than $750 per average kilowatt in keeping with the General Electric estimate for a production run of only 100 units, as shown in table 2, rather than to take into account also the earlier estimates as we have done. With the transmission line cost estimate also boosted by 25 percent this would come to $1,250 per average kilowatt for wind power delivered to an urban electric grid four hundred miles away. Our estimate as already explained is that wind power with four hundred miles of extra transmission line but without storage costs $1,000 per average kilowatt, only 40 percent of the cost of nuclear power without storage at more than $2,500 per average kilowatt. By stretching the estimates in deference to possible skeptics by 20 percent in favor of nuclear and 25 percent against wind power we come out with $1,250 as compared with $1,900, that is, with the capital cost of wind power thus delivered about two-thirds of the capital plus fuel cost of nuclear power. Without the extra transmission line, the wind-power figure thus stretched is still only half of the nuclear one.

At least for the first few thousand large wind dynamos in windy regions, these last figures are the ones that confirmed skeptics should be thinking of, for there are many uses not requiring storage, either supplementing hydroelectric installations or pumping irrigation water or supplying nearby special industries or else supplementing municipal electric grids with a few percent of their power. Although intermittent, the municipal load tends to peak in cold windy weather when the electric heating load is greatest.

If we include also three-day storage in the wind-power system, with its cost also stretched by 25 percent, so that the wind power can tide over a week or so of light winds and also have the capability of meeting daily peak demands, which the nuclear system without storage cannot meet, the skeptic's stretched figure for wind power is 90 percent of that for nuclear power. Without stretching the figure to satisfy skeptics, out straightforward estimate for this ratio is 1450/2600, or 56 percent.

10

Social Costs of Wind Power
and Nuclear Power Compared

In the course of a decision on the extent to promote and adopt various power sources, dollar costs figure in two different ways. The direct dollar costs considered in the previous chapter are those borne by the private investor and influence his consideration of the prospects for profit to repay the investment. There are indirect costs borne by the public through government channels, either purposefully for the sake of stimulating private investment and furthering the use of the power source, or inadvertantly without really appreciating what the costs may be. Since these costs are paid by society in general rather than specifically by the consumer of the power, these may be classed as social costs.

Other and perhaps more important social costs are those borne by society in the form of degradation of health and life and the environment (Commoner, 1975). Both types of social costs should be seriously considered and weigh heavily in the establishment of public power policy. Even when there is a perceived need for a change it may be politically very difficult to alter the direction of policy because the technical expertise and invested capital in the current commercial enterprise have great influence in policy decisions. This makes it all the more important for the public to appreciate what the social costs may be.

Social and Environmental Costs of Large-Scale Wind Power

Siting and Visual Impact of Wind Turbines
The massive, old-fashioned windmills of Western Europe, particularly of Holland, have long been considered picturesque and some of them remain as tourist attractions, as do a few in this country. Although very large modern wind turbines with rapidly rotating, slender, propeller-like blades will be less conspicuous than the old-

178

fashioned ones, they will be taller and the prospect of large numbers of them on scenic mountaintops or in densely populated areas is not an appealing one to most people. Pictures of early Paris show windmills on the city walls and within them, tolerated by city folk. A functional engineering design of metal girders is less appealing than the sturdy old windmills that have antiquarian value, yet Paris is still proud of its Eiffel Tower, built a century ago of steel girders and loftier than even very large wind turbines.

Sometimes the term "visual pollution" is used to describe unsightly intrusions of technology on the landscape, like large windmills and the cooling towers of steam power plants. It seems unnecessary and unlikely that large windmills will ever be common in cities or suburbs, though small wind dynamos may take their places with television antennae, tolerated as mild visual pollution of the urban scene. With slender blades and a shapely nacelle atop a single pole, a small wind dynamo may even be considered attractive. Larger ones may appear in industrial areas where aesthetic limitations have already been broached.

But for large-scale electric network power generation there need be no visual objection to large wind dynamos largely because economic considerations will put them where the winds are strongest and steadiest, and these tend to be in sparsely populated regions. Much of the area of the windiest part of the western Great Plains is arid grazing land with few head of cattle per square mile. A similar low density of giant wind turbines would not seriously interfere. Some of the land is intensively farmed in circles half a mile in diameter irrigated with pumped water sprayed from long, slowly rotating radial pipes. The unused corners between the circles would make good sites for large wind dynamos. A small part of their output could be used locally to pump the irrigation water. A widespread offshore array of floating and moored wind dynamos should not seriously interfere with the activities of commercial fishermen, and might even be welcomed as convenient navigational fixes, like buoys.

It may be that mountaintop sites will prove to be economically attractive, the advantage of the higher winds more than compensating for the added costs of access facilities as was the case in the Grandpa's Knob experience. If so, it should be possible to confine exploitation of such sites to remote ones seldom visited for scenic or recreational enjoyment, such as the treeless lesser mountains in the arid intermountain reaches of Nevada. Most of the choicest mountain ranges are on public land where this exclusion could be made a matter of policy. However, within the prospect of deploying thousands of large wind dynamos to supply a substantial part of our electric power needs, it seems unlikely that mountaintop locations will be used because each mountain location is an individual siting problem. There would be a preference for the uniformity of siting that is available for large arrays of machines in the windy regions of the Great Plains, even though there may be gullies and arroyos to break the pattern of an array.

Thus the visual impact of wind turbines that is sometimes cited as an objection to wind power by people thinking they would be sited where many people will see them need not be an objection to the large-scale use of wind turbines for commercial electric power generation.

Storage and Transmission Requirements of Wind Power
Both wind power and fuel-consuming power sources can benefit from the use of pumped hydroelectric storage, as we have seen. Under most circumstances wind power is somewhat more dependent on storage because of the unsteadiness of the wind and the environmental disadvantages of pumped hydroelectric storage apply more to wind power than to fuel-consuming plants. As with conventional hydroelectric power, land occupancy is one of the environmental costs of above-ground pumped storage. Others are fluctuating water levels and interference with aquatic life if a river is used as a lower reservoir. A mountaintop reservoir such as the one at Northfield, Massachusetts, is inconspicuous and displaces only forest much smaller in area than the prime river-bottom land flooded by many reservoirs for city water supplies. Future underground pumped storage would not have even this small environmental impact.

Practically all sources of commercial electric power require above-ground transmission lines, a form of visual pollution to which we have become accustomed. Wind power generated in remote windy regions will require longer stretches of lines than do fuel-consuming plants that can be located near cities. Both require transmission lines in the densely populated regions near cities but the additional lines required by remote wind power are mainly in the more sparsely settled countryside, such as the farmlands between western Kansas and Chicago, where they bother fewer people. There is pressure for building a nationwide power grid involving these long lines for other reasons, into which shorter feeder lines would be required for wind power.

Effect on Bird Life
A rapidly rotating wind turbine blade could kill or injure a bird flying through the blade circle, though it is very unlikely that a bird,

even if it does fly through the circle, would be hit by a blade. The blades of a wind turbine twist their way through the passing air-stream like a giant corkscrew. The thickness of the slab of wind between two successive passages of a blade, if there are two blades with a tip speed seven times wind speed, is $\pi R/7$ or about half of the radius of the blade. For a large wind dynamo this is about fif-teen meters or fifty feet, a very thick slab in which the bird can fly without being in the path of a blade. Even if it should happen to be directly in the path of a blade, it may be deflected around in the airstream and merely graze the surface, because at least a small bird is very light and moves with the air.

Information about birds being killed by windmills seems to be nonexistent. Inquiry in Denmark and of Professor John Wilber of MIT, managing engineer of the Smith-Putnam project at Grandpa's Knob and now retired in Hancock, New Hampshire, offers no recollection of observation of birds having been killed. Birds in-stinctively shy away from moving objects and even a fast-moving blade is visible and would probably turn a bird in flight away from the wind turbine in daytime. A wind turbine would serve as a giant animated scarecrow. The swishing sound of the blade could have a similar effect at night. Even if there should be a problem, some kind of augmentation of sight or sound could probably be devised. As for seabirds tangling with offshore wind dynamos, they do swarm behind fishing boats but would have no reason to seek out wind turbines. It is hard to imagine a toll comparable to that from oil spills and wind power would reduce oil transport. On land, there is special concern for the safety of whooping cranes, very rare large birds, but their flyway is close to the Mississippi River where they are in more danger of potshots from uninformed river-men. Their routes are hundreds of miles east of the windy western plains where wind dynamos should be concentrated. The sky-scrapers of New York City do take a toll of songbirds migrating at night down the narrow coastal flyway, but those buildings present very large areas compared with slender wind turbines and make no warning sound. So far as one can tell, for wind turbines this is not a problem.

Possible Effect on Weather Patterns

To every action there is an opposite reaction: as the wind drives a wind turbine, the wind turbine slows down the wind. The question arises whether, if a very large array of windmills were spread thinly over a large area, this could have any appreciable effect on the future behavior of the wind, and in particular whether it could in any way modify the weather.

So little is known about this effect that it should in the first place be remarked that, if there should be a noticeable effect, one cannot tell in advance whether it would be an improvement in the weather or make it worse. It may depend on one's point of view— no one considers the present weather pattern perfect.

One reason to conclude that a huge array of wind turbines supplying half or even more of our electric power, 10 percent or more of our total power consumption, could have no appreciable effect on macro weather patterns is that the energy content of wind motion at any instant is as great as mankind's rate of use of energy at the present rate in five years (see appendix 1). Yet the power of sunshine, even working through the inefficient heat engine of the atmosphere with its wasteful radiations, could replenish that energy in a couple of days.

The main effect of a large array of wind turbines on the circulation of the atmosphere is to increase slightly the frictional contact between the lower layer of the atmosphere and the ground. Force is associated with a change of momentum, and the force exerted on the wind turbines implies that there is a downward flow of momentum in the layers of air just above them, arising from increased shear and viscous friction as well as upward and downward motions of the air. Thus the wind turbines are extracting energy and momentum from a layer of air considerably thicker than their height.

A very large array in the western Great Plains would consist of wind turbines perhaps two hundred feet high spaced half a mile apart, which is a sparse coverage of the land. The effect on the wind would be much less than if big trees were planted there a hundred feet apart, and still less than if a dense forest grew there. Such an experiment is hard to carry out but an opposite experiment has been carried out, with no known effect on the weather, namely, in the deforestation of much of the northern states in the Northeast. New England was deforested two centuries ago and has mostly grown back to forest as agriculture moved west in the last century. Since that is a hilly and mountainous region, it is perhaps more impressive that the flat north central states, particularly Michigan and Wisconsin, were denuded of their virgin forests by a thriving lumber industry that helped build the country a century ago. They lay almost bare for some decades before gradually becoming forested again with smaller trees. No one seems to have traced changes in the weather to this huge operation, yet one would think it would have more effect than sparsely covering comparable parts of the flat western plains with wind turbines.

There are gradual natural changes in climate and average weather, so it is hard to detect changes caused by man's activities. Even where theory may suggest a small effect from man's activity, mathematical modeling of the atmosphere has not yet reached the stage of being able to determine whether such relatively small influences as deforestation would slightly accelerate or retard the natural trend. Articles in the interesting compendium *Man's Impact on Climate* (Matthews, 1971) discuss many possible influences by man, such as carbon dioxide, dust, waste heat, melting of ice caps, aerosols, jet exhausts, and even oxygen production by marine microorganisms affected by chemical pollution, all as possible inputs to mathematical modeling; yet the only influence of large-scale deforestation considered is the change in reflectivity of the earth's surface, resulting in a net cooling effect because forests absorb sunlight. The influence of deforestation on surface wind drag, which might be relevant to windmills, is not even mentioned.

In more ancient examples of deforestation there is thought to be evidence that deforestation and some of its consequences do affect the weather and bring on drought; striking examples are found in the Sahara Desert and in the Middle East where desert has replaced a veritable Garden of Eden and great cities lie abandoned. If this evidence is accepted, it would seem that any small change a sparse coverage of wind turbines might make would be in the direction of increasing rainfall at places where it is much needed. Even if this should thereby slightly decrease rainfall in some places where there is already more rain and at sea where rain is useless, this would be a net gain. It does seem plausible that an increase of ground-layer drag, by encouraging atmospheric mixing with upward and downward drafts and thus cloud formation, might have a slight effect increasing local rainfall. All in all, then, the possibility of weather modification cannot be counted as a hidden cost of wind power.

Economic Stimuli to Industry
The federal budget for research, development, and demonstration of wind power started at one-fifth of a million dollars in 1972 and has rapidly grown to $37 million in fiscal 1978. In addition to this social dollar cost which is expected to continue into the middle of the next decade, it is proposed in the following chapter that direct economic stimuli should be paid from the public purse, either directly or through loss of revenue, to help get industry through the economically difficult stage of achieving mass produc-

tion of wind dynamos quickly. This would help counterbalance similar stimuli that have been, and continue to be, supplied to competing energy sources and are considered part of their social costs.

Congress appropriates billions of dollars in an effort to reduce unemployment. Though they need not be that large, appropriations to stimulate mass production and deployment of large wind dynamos, a labor-intensive industry, could well be considered a part of that effort.

Social and Environmental Costs of Nuclear Power

The dollar costs of nuclear power discussed in the previous chapter are the direct ones that must be considered by utility executives in deciding whether to invest company capital in nuclear power. There are other dollar costs borne by society, through government spending and taxes, that will not show on company balance sheets. There are also further costs borne by society not in the form of dollars but mainly as risks to health and life and degradation of the environment. Of course, trying to cope with these involves dollars too.

Some of the social costs have been discussed already in chapter 7 as disadvantages of nuclear power. That discussion is supplemented here to bring them explicitly into the reckoning of the total cost. On his balance sheet the businessman may ignore them, aside from the items supporting lobbying and advertising to assure that they continue to be borne by the public. The powerful influence thus exerted makes a truly public decision difficult, but ideally the public decision should balance against the benefits to be derived from the power the total cost, direct and indirect, dollar and social and environmental. At the very least, we should recognize what these costs are. The benefits should be assessed only to the extent that they surpass those of the alternatives, and this is where conservation and wind power, among others, come into the reckoning in a comparison of two cost-benefit analyses.

Development Costs
Nuclear power plants, first used for submarine propulsion and later for civilian electric power, are a spin-off from the development of reactors at first completely, then partially, devoted to the production of plutonium for military purposes. There is thus no sure way to separate the military and civilian development costs. Until the mid-fifties when the "Atoms for Peace" program was launched, the costs might generously all be charged to the mili-

tary, though most of the conceptual design and pilot-plant development of present commercial reactors was done largely in national laboratories with civilian-development motivation; and the commercialization of nuclear power could be started then only because of the existence of expensive isotope separation facilities built during World War II. A great new technology was given to industry on a silver platter, so to speak. During the following decade, one of the arrangements for subsidizing commercial nuclear power was for the government to lease the prepared nuclear fuel to the power companies and to pay them for the plutonium separated out from the returned fuel, to be used for bombs. The power plants thus had dual-purpose reactors. This arrangement was continued beyond the military need for the sake of continuing the subsidy, the plutonium being stockpiled for possible future use in power reactors. Here again the line between military and civilian costs cannot be defined.

During the rapid bomb-stockpile expansion in the early sixties, sparked by the mythical "missile gap" of the 1960 election, multi-billion-dollar isotope separation facilities were added, supplementing the wartime Oak Ridge plant, to meet both the military objective and the foreseen civilian need for reactors then under construction or planned. This peacetime expenditure should probably be charged half to the civilian power program, for which it has been essential. If a reckoning were available of these costs added to similarly apportioned costs of maintaining the national laboratories to provide the continuing research backup of industrial reactor design and safety measures, along with the expenditures and lack of revenue through tax write-offs to encourage prospecting and development of uranium sources, such a reckoning would undoubtedly come to quite a few billions of dollars by now. If there were, say, 10 billion 1977 dollars, to be apportioned to the approximately 50 gigawatts of present commercial nuclear generating capacity, this would be about $200 per kilowatt. Most of those plants were built early enough to be priced at less than $400 (1977) per kilowatt, so the public contribution via the government appears, from this estimate, to have been more than half of the direct cost of the plants. In fact, looking at the figures just up to 1975, the government had spent more than 8 billion dollars (in 1975 dollars) on civilian nuclear power, amounting to something like 85 percent of the capital that industry had invested in all the commercial power plants that were then in operation (Welch, 1977). What the government pays comes, of course, partly from taxes and partly from minting new money that contributes to inflation.

Thus the costs that the public bears in this way for the services provided by government to make the nuclear industry seem competitive are very substantial. Those are costs already paid, about which no further decision can be made, water over the dam but indicative of the way the wind blows.

If we accept the development of our present nuclear technical and fuel-supply capabilities as a gift from the past, beyond further decision, the present consideration of how much we should depend on nuclear energy in the future should recognize the present cost of preparations for that future, which may be considered part of the present cost of nuclear energy.

The federal energy research and development budget outlays proposed for fiscal 1977 were as follows (in millions of dollars, ERDA, 1976).

Nuclear fuel cycle and safeguards	306	} 1,095	} 1,399
Fission (mostly the breeder)	789		
Fusion	304		
Conservation	93		
Geothermal	56		
Solar	117		
Fossil	524		

Thus the federal expenditures for promoting the future of fission technology and fuel supply, including the development of the LMFBR, the breeder reactor, are just over a billion dollars annually, with an additional 0.3 billion dollars annually pinned (actually gambled) on the hopes for future commercialization of fusion energy. Such budget figures have been increasing from year to year, in constant dollars. This expenditure of about 1.4 billion dollars a year gives the currently operating nuclear power plants the hoped-for status of being part of a continuing energy supply program rather than being on a dead-end street, at an equivalent cost of about 35 million dollars a year per 1,000-megawatt plant.

As for the near future, these costs are rapidly mounting and it is expected that by the time a commercially viable breeder may have been developed, the government will have put about 14 billion dollars (in 1977 dollars), into it, up from about 4 billion dollars so far. In view of this and other announced obligations and extrapolations from present budgets, B. L. Welch (1977) estimates that within the next decade federal support for the civilian nuclear power industry will easily exceed 25 billion dollars (in 1975 dollars), with half as much again spent by the TVA, the world's largest purchaser of nuclear power plants. He adds prophetically, "One hidden cost may well be our free enterprise system. Clearly, nuclear fission is an expensive way to boil water."

Tax-Incentive Loss of Revenue

For the sake of promoting industry and our competitive position in international trade, energy has been made artificially cheap in this country by providing a remarkable tax incentive to people investing in the extraction of fossil fuels and minerals like uranium. The tax provision is referred to as a depletion allowance, suggesting that it is treating uranium deposits similarly to the way one can allow for the depreciation of the value of a house from wear, but this is misleading. Actually, it depends not on depletion of the source but on sales price of the product. In figuring income tax, an investor is permitted to deduct 22 percent of the sales price (not just the profit) of uranium he sells not just from his uranium income, but essentially from his net income from all sources, up to a limit of 65 percent of the total. This is the sort of arrangement that makes it possible for some people with large incomes to pay relatively small taxes and it makes investment in uranium mining attractive enough to stimulate the supply and help keep down the price of uranium. But it does cost the public money. What those investors do not pay in taxes, the rest of the public does. This is one of the hidden dollar costs of nuclear power carried by society.

Decommissioning

It is anticipated that nuclear power plants will have a useful life of about thirty years. No provision is made in present funding for what happens to the plant after that. It is not like an ordinary power plant or factory that can be simply torn down or converted into something else, because even after the spent fuel is routinely removed, the reactor itself and its immediate surroundings will be radioactive as a result of deposition and neutron bombardment over all those years. The Navy encountered an unexpectedly formidable task in decommissioning a small power reactor at McMurdo Sound in the Antarctic, which portends the difficulties and expenses that may be incurred when the time comes either to remove spent nuclear power reactors or to leave them as permanent features of the landscape. Rationally, each nuclear power plant should be required to contribute regularly from its revenues to a decommissioning fund as part of the present cost of nuclear power. As it is, the public will presumably have to meet this future expense when the time comes.

The world's first nuclear reactor, the small graphite research reactor at Chicago, after having been reassembled in a park outside the city and operated for a decade, was decommissioned in about 1954 in order to return the park to its original use. A pit was bulldozed next to it and it was shoved in and buried. Such simple treatment will not do for a large and more radioactive modern power reactor. If the core is to be left in place, it is considered necessary to surround it with concrete two meters or more thick as a gamma-ray shield unless people can be dependably excluded from the area for many years. The pressure tank is also radioactive and it can be disposed of by burial after flooding the whole reactor with water and cutting the thick metal with underwater welding torch techniques, but this is very expensive. The Elk River reactor in Minnesota, fairly small by modern standards, was decommissioned in this way at a cost greater than the cost of building it in the first place.

A more reasonable solution would be to require that the reactors be built underground in the first place. Decommissioning would then leave them buried by merely blocking the access ports. This might be an expensive provision if it were for decommissioning alone, but the decommissioning convenience would be a side benefit, the main purpose of underground construction being stronger containment as protection against a catastrophic accident (Watson, 1972; Inglis and Ringo, 1957). Estimates of the added cost of putting a nuclear power plant underground, based partly on European experience in doing so, have been in the neighborhood of 5 percent to 10 percent of the cost of the plant; however, a recent ERDA-sponsored estimate is about 40 percent. Ardor for promoting apparently cheap nuclear power has not permitted either this or the foresighted requirement of a decommissioning fund so this social cost remains to be paid in the future.

Liability Insurance
Owners of normal fossil-fueled power plants, like automobile owners, carry liability insurance in case something goes wrong, such as a boiler bursting, and causes injury. The potential damage is limited and the insurance, while expensive, is not prohibitive. It spreads out the possible consequences of the risks among many owners at a considerable cost in overhead.

The consequences of a catastrophic accident of a nuclear power plant are essentially unlimited, though the greater the damage of a hypothetical accident, the smaller the likelihood generally is that it may occur. When, in 1957, the first reactor safety study report mentioned seven billion dollars as the possible property damage of a hypothetical nuclear accident, consortiums of insurance companies were willing to sell insurance to cover only a few percent of that. This difficulty in the government's promotion of nuclear power was met by prompt passage of the Price-Anderson Act (see appendix 3, p. 251) which stipulates that a reactor owner must carry commercial insurance for liability up to about one hundred million dollars, that the government will supply the additional insurance up to about half a billion dollars, and that there shall be no liability for any damage greater than that. While damage less than this limit is presumably more likely than greater damage, this means that the public simply takes the financial risk, without recourse to compensation, over by far the greater part of the range of possible damage.

During the hasty 1975 congressional debate on renewing the Price-Anderson act for another ten years, it was pointed out by nuclear proponents that the act had not cost the government a penny in compensation benefits, making it sound as if this arrangement costs nothing. Similarly, insuring your auto costs the insurance company nothing but overhead until you have your accident, but you pay premiums nevertheless. All it means is that the nuclear accident covered by the insurance hasn't yet happened. When and if it does, the government tax-plus-inflation revenue base, meaning society, will pay its part and the uninsured public will have to accept any cost there may be beyond that.

Radioactive Release and Radioactive Wastes
The items of cost listed above are those that manifest themselves primarily as incidental financial costs of nuclear power borne by society, expressible in dollars. Now we come to those whose primary impact and cause for concern is in the realm of human health and the environment, the health and environmental costs.

In the section on insurance we discussed the financial aspects of the risk of calamitous nuclear accident. The human aspects of that same risk may be considered more serious than the financial aspects, for most people would like to value life and health above money. Short of coldly putting a dollar value on human life and health, one can compare the costs of various power sources in two separate ways, the dollar costs and the human-plus-environmental costs. Then it becomes a matter of personal judgment to weigh the two separate cost comparisons in the final choice of options. Just as in the case of the financial risk, the human risk should be counted as part of the present cost of nuclear power even though a catastrophic nuclear accident has not yet occurred, and may not within the span of the nuclear power program. It is an actuarial risk, hoped to be a small probability of occurrence of very undesirable consequences that might happen at any time.

The risk of calamitous accident is perhaps impossible to quantify credibly ahead of the event. The latest monumental governmental attempt to do so will probably be judged, by most people who take the time to look into it carefully, to have been motivated and used largely for promotion purposes, and to be so deficient in its methodology and coverage of sources of trouble that its evaluation of the risks has essentially no meaning. This situation is outlined in chapter 7 and appendix 3. Though that study is often quoted as proof that there is almost no risk, one may come away from it with the feeling that the only reliable limitation we can put on the estimated risk of disastrous accident comes from the fact that one has not yet occurred in about two hundred reactor-years of commercial power plant experience. This fact is indeed often quoted by utility companies as proof of nuclear safety. With only this and the incidences of near misses within those two hundred reactor-years as a basis for judgment, a pessimist might, without being an alarmist, conclude that with a thousand big reactors operating in the future there might be an accident killing hundreds of people with prompt or slow radiation effects about once a year, while an optimist may conclude that, since the near misses have been successfully stopped short of calamity, this will always be so and there will be no catastrophes. With all its uncertainty, the risk of calamitous accident is a serious but unquantifiable human cost of nuclear power.

Radioactive releases from normal operations are another human cost of nuclear power that cannot yet be quantified with confidence. As the discussion in chapter 7 suggests, the difficulty here is in establishing confidence in the assumptions it is necessary to make on the basis of inconclusive data, about the effects of low-level radiation. Releases from the entire fuel cycle including mine tailings must be considered and if present regulatory practices are followed, the direct release from nuclear power plants may be a small part of it. When fuel was being reprocessed, those releases were considerably greater than from power plants and the extent of the total release from the fuel cycle can be suppressed by continuing a once-through use of uranium resources, leaving the plutonium and most of the fission products locked in the fuel pellets for perpetual storage. The stored spent fuel, being a chemical mixture in solid form, would not be easily dispersed and would require special handling in chemical processing to be available for malicious dispersal or bomb-making. This once-through practice has been proposed also in connection with the proliferation problem, as is discussed elsewhere. It would translate these concerns for human costs also into dollar costs.

The production of plutonium primarily for nuclear weapons has already created a monumental problem of radioactive waste disposal that will only be made worse if we go ahead with the nuclear power program, particularly in view of the potential huge plutonium economy based on the Liquid Metal Fast Breeder Reactor, as has been mentioned in chapter 7. Search for a permanent solution of the problem during the more than thirty years of reactor development has yielded none sufficiently satisfactory to have yet been instituted and the nuclear power program goes on generating more wastes on the assumption that one eventually will be found and adopted. Some wastes are being reduced to solid form but are not yet permanently sequestered. When and if a satisfactory underground permanent disposal site is found, unless it is so remotely deep as to be henceforth inaccessible to man it will have to be protected against accidental or malicious intrusion; quite a nuisance for future generations involving some uncertainty of effectiveness at times of political upheaval or natural disaster. The hope has been expressed by ERDA spokesmen that the current dollar cost of disposal can be kept down to not much more than a tenth of a percent of nuclear power costs, but this seems very optimistic in view of an estimate that disposal of the present burden of wastes may cost the government tens of billions of dollars if disposal sites can be found (L. J. Carter, 1977). The responsibility for assuring the security of sequestered waste is a legacy that we pass on to future generations as part of the social cost of our present nuclear power. This is of course not the only legacy of our use of nuclear power that is both serving current demands and setting the stage for the future. Perhaps, for example, we will pass on a pattern of excessive use of power that cannot be sustained but will be politically difficult to terminate.

Proliferation of Nuclear Materials

As we have seen, the promotion of the worldwide use of nuclear power spreads the availability of plutonium and facilitates the making of nuclear bombs by various groups around the world, increasing not only the threat of nuclear terrorism but, far worse, the very real risk of nuclear war that most people prefer to ignore. In comparing the social costs of nuclear and wind power, the contribution of nuclear power to the risk of nuclear war is perhaps the most compelling reason for instituting a crash program of building and using large wind dynamos, both to set the example for others to follow and to substitute wind power, and ultimately other solar sources, for nuclear power exported to developing countries.

The proliferation of nuclear materials is a disturbing cost of nuclear power itself and of the high priority we are giving it over other technologies. That this program should have been inaugurated under the banner "Atoms for Peace" is prominent among the delusions that have been spread about nuclear power. One might better say, "Wind Power for National Security" (Brown, 1977; Gruener, 1978).

11

Choice of Options

Prospects for Growth

There long has been and still is contention over the extent to which the globe can support continued rapid growth in human population and energy-consuming activities. Malthus is famous for having incorrectly predicted in 1798 that England would be at the end of its economic rope in a decade or two. He apparently did not appreciate that England was not living in isolation on the resources of a small island, but rather was the center of an expanding empire drawing on much of the world for its material well-being. His mistake is sometimes taken to discredit those who today see the expanding needs of humanity on a collision course with the globe's limited resources. Optimistic analysts foresee an incredibly rich future world population, several times as numerous as now (Kahn, 1977). Some of them even take seriously the prospect of space colonization as projecting human population into interplanetary space and giving Earth the benefit of having an expanding colonial empire, much as England had two centuries ago. Other analysts see the period of rapid population and material growth of the recent past, even with all its troubles and suffering, as a time of exceptional good fortune with favorable climate, rapid exploitation of mineral resources, and a green revolution that has not yet run its course. Prompt curtailment of growth, most importantly of world population and, second, of consumption of expendible energy sources, is seen as the road to maximum human happiness and minimum suffering.

But no matter which of these trends is followed, whether total energy consumption continues to rise sharply or whether it soon levels off as it seems to have begun to do, prompt introduction of large-scale wind generated electric power is urgently needed. With continued rapid energy growth most of the practical sources will be needed, and the more wind power the better. With modest growth wind power should soon begin to replace more exhaustible and more damaging sources.

Until recently it was generally expected that growth would continue on an exponential rise with a fixed doubling time of about ten years for electric power consumption in the United States and about twenty years (or an annual growth rate of 3.4 percent) for consumption of all types of energy, perhaps faster. Since 1973 the administration's expectation has been revised downward but not very drastically, leaving the anticipation of continued rapid growth.

A far-ranging study by the Energy Policy Project (Ford, 1974) has outlined three possible scenarios for growth to illustrate possible trends of our future economy.

1. The *Historical Growth Scenario* continues energy consumption at a rate of 3.4 percent, much of the energy wasted in producing and converting energy inefficiently with no margin for selectivity of the means of production.

2. The *Technical Fix Scenario* eliminates the waste of both energy production and consumption by greater attention to technical improvements and accomplishes just as rapid a growth in the economy and the amenities of life with the annual rate of growth of energy consumption reduced to about 1.9 percent for the rest of the century and continued growth beyond.

3. The so-called *Zero Growth Scenario* anticipates continued growth of energy consumption only slightly less than in (2) for the next decade or so but tapering off to zero growth rate by the end of the century. This uses the technical fixes and anticipates about the same growth of gross national product as do the other scenarios but growth based on some change in life style emphasizing more services, fewer and more durable goods, and less damage to the environment.

The energy growth curves for the three scenarios are compared in figure 46. It is to be emphasized that all three scenarios include about the same overall economic growth. The marked differences in energy consumption arise from differences in efficiency and in the distribution of enterprise between goods and services.

Each of these scenarios was seen in the report as giving us a satisfactory standard of living, the first a wasteful one leading rapidly to depletion of resources, the latter requiring some change in our general living patterns with less emphasis on suburban expansion into large living areas and a prompt introduction of economies in transportation and building practices, requiring architects to become concerned about energy conservation, but no real austerity.

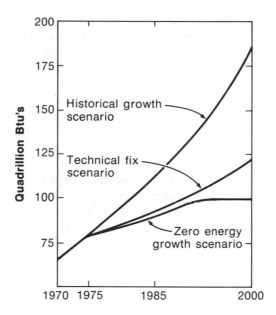

Fig. 46. Scenarios: Energy use in 1985 and 2000. (*Courtesy Energy Policy Project, Ford Foundation.*)

While it might be desirable, it seems unlikely that a forthright decision will be made selecting between these options. Rather the trend will be set by a large number of decisions influenced by economic and socio-political pressures in a number of fields. The availability and choice of energy options will both be one of the pressures influencing the trend and will be influenced by the trend established in response to other influences.

There are thus two interdependent kinds of options to be considered, the option for the trend of the economy along the lines of one or another of the three scenarios and the option between energy sources. We may in either case distinguish between the ideal or socially desirable option and the likely or politically practical option. Although some industrial leaders would still choose the first scenario, something like the third scenario is considered by many to be the ideal option, having superior long-term prospects leading to a stable world economy.

Amory Lovins (1975, 1976, 1977; Barney, 1977) goes further and makes a good case for leveling off energy consumption and particularly electric generating capacity more rapidly than in the third scenario, reaching a maximum in perhaps a decade and shrinking to a somewhat lower steady level. He emphasizes the fact that only a modest part of our electric energy is used for purposes to which electricity is uniquely suited, such as subways and elevators, and smelting and the household conveniences aside from space heating. Much of the rest is used inefficiently doing things that could be done better by what he calls "soft" technologies that could well supply an increasing share of the shrinking electric load as the hard sources more demanding of natural resources are gradually retired.

Attractive as his analysis of such prospects appears, it seems unlikely that we will change our ways so fast and that the decisions concerning adoption of wind power will more likely be made in the context of a growth pattern close to the third scenario or somewhere between the second and third, either of which envisages continued growth in electric generating capacity for some time to come (Hammond, 1977b). The sooner growth in electric generating capacity levels off, the sooner it will be possible to start replacing existing capacity with wind power and other solar sources rather than just reducing the need to build new fuel-consuming plants.

These scenarios are useful mainly in indicating the wide range of future energy conservation trends, depending not only on administrative decisions making a choice of options but also on various influences that may defy the administration's intentions. Just after these scenarios were described, the historic growth of energy comsumption in the United States suddenly broke off following the drastic increase in fuel prices in 1973. The consumption remained somewhat less than the 1973 level for the next three years, mainly as a result of conservation, eliminating some of the most flagrant and most easily modified practices of wasting energy, under pressure of higher energy cost. This could be viewed as a one-time reduction below the rising curves of the three scenarios and resumption of some increase is to be expected at a rate somewhere within the range depicted by the three scenarios.

Energy Policy

The Federal Energy Plan prepared by the Carter administration early in 1977 is based on the anticipation that without any further planning the growth in energy consumption would continue only slightly less rapidly than in the Historic Growth Scenario, increasing by about 30 percent as compared with 38 percent in the Historic Growth Scenario or 12 percent in the Zero Growth Scenario, for the nine-year period between 1976 and 1985. Despite President Carter's statements that he would rely mainly on conservation, the Federal Energy Plan would reduce energy consumption only slightly below the no-planning figure, from 30 percent to about 25 percent (Berger, 1977). Whereas conservation measures might be expected to reduce per capita energy consumption, the new plan would increase it by about 20 percent in nine years and the average individual American will go on consuming more and more energy to meet his presumed needs. Political pressures for business as usual are so great that it seems unlikely that even this modest goal of the plan will be achieved.

The main objective of the Federal Energy Plan is to reduce dependence on oil imports in the near future. It proposes to achieve this by greatly speeding up the rate of depletion of United States fossil-fuel reserves (including those of oil needed for future strength) and at the same time accelerating the construction of uranium-consuming nuclear power plants. President Carter promised to use nuclear power "only as a last resort." Secretary Schlesinger, while continuing a policy favoring nuclear power over solar-related alternatives, somewhat cynically remarked that "there will be a lot of the last resort." It seems to be planned that there will be enough of it to increase the nuclear generating capacity of the country three- or four-fold in the next decade, which will mean consuming a lot of energy for construction in the meantime. In fostering rapid depletion of our oil and gas reserves in the short term, it increases the prospective need for renewable resources such as wind power in the long term and yet makes no adequate provision for starting their large-scale use soon to give them time to grow.

The Federal Energy Plan as presented early in the Carter administration need not be the last word and will probably evolve in various ways. Its stated goals are mainly for 1985 and beyond, expecting to require much less foreign oil after that date but more in the meantime, along with increasing amounts of domestic coal and oil. How the growth is to be sustained in these next ten years is not completely clear. It usually takes about ten years to plan and build a nuclear power plant so only nuclear plants already planned and under construction will contribute appreciably in that time. Fossil-fuel-fired plants come on line more quickly so they will probably constitute most of the growth. However, with a sufficiently forthright decision, wind power could also help substantially in the latter half of the decade, reducing the number of other plants needed. The fossil-fuel plants each have undesirable characteristics, coal being dirty and oil and gas being in short supply, and much more valuable as future chemical feedstock than as fuel, so one would like to think of these plants as temporary for meeting the needs of the next decade, to be retired as soon as possible thereafter as other technologies become available.

It is officially anticipated that wind power along with other solar sources may play an important role after the turn of the century. It would be much better to speed things up and do it right in the first place, avoiding making the full commitment of capital and materials to the fuel-consuming technologies that will make it still harder to introduce the ecologically preferable sources later on. Of wind power, the one of the solar sources that is already developed and ready to go, we may ask "eventually, why not now?" Getting it started rapidly enough to make a substantial contribution during the latter half of the next decade would of course facilitate its further expansion after that.

How rapidly wind power can get started will depend on whether we make a virtue of letting it be done by individual enterprise in the private sector, unaided by government incentives after the initial demonstration phase, or whether we give it the advantage of subsidies and tax advantages such as have benefited the other energy sources with which wind power must compete. As a report of the Stanford Research Institute says, "A nation that wanted to ensure against foreclosing the future options would make a social decision to implement solar technology far more rapidly than economic decision making would otherwise warrant."

The present wind-power program counts on "economic decision making" to determine the rate of expansion after the demonstration and testing contracts come to an end. To quote from *A National Plan for Energy RD&D* (ERDA, 1976, vol. 1, p. 105), "The principal federal role is to assist the private sector in the development and improvement of wind energy conversion technology, and thereby provide a stimulus for private industry to produce such systems and for utilities and others to use them in suitable applications."

Even without any additional incentives, the program is expected to make a modest contribution toward the end of the next ten years. To quote further, "Near Term: (—1985). It is estimated that a successful RD&D program whose results would be implemented at an early date by industry could be capable of supporting commercial energy production of 2.5 to 5 10^9 kW$_e$H per year . . . " that is, about 300 to 600 megawatts on the average. By the year 2000 it is anticipated it could be 20 to 35 thousand megawatts, the equivalent of about 33 to 60 large (1,000-megawatt) steam plants at 60 percent capacity factor. That amounts to about 8 to 14 percent of our present national use of electrical power or roughly 1 percent of our present total use of power.

This in keeping with statements that high officials have been making for several years to the effect that wind power cannot be expected to meet more than 1 percent of our power needs before the end of the century. It shows that these statements are based on the assumption that industry is to be left to pick up the ball and run with it after having been helped by ERDA and DOE through the demonstration stage. If a social decision should be made to implement wind power more rapidly, there are several ways for the government to make the demonstration stage more effective and to interject economic stimuli to accelerate the implementation of large-scale wind power, as will be discussed further on.

A social decision to change policy can be difficult to achieve when established industries are involved. As a background to considering the possibility of a change, let us consider how the present emphasis of our energy policy became established, and in particular the overwhelmingly preponderant role assigned to nuclear power as the supplement to fossil fuels in the period beyond the next ten years, with only a minor part slated for wind power and other solar-related sources until sometime in the next century.

Evolution of Energy Policy
The very disparate emphasis on nuclear and wind power is a natural outgrowth of the history of government involvement in power production. No unbiased rational decision has been made choosing between the energy options as they present themselves now. It is time that such a decision should be made before we automatically go into the next round of huge expenditures to help nuclear power out of its current difficulties.

The whole matter was essentially decided back in the early 1950s. The decision to go into government sponsorship of nuclear generation of electric power was made as a natural extension of the wartime bomb effort and the subsequent nuclear submarine development, as is outlined in chapter 6. Since then the growing nuclear emphasis has been a matter not of selecting between options but merely of drifting, of responding to the inexorable pressures of established institutions.

In this connection it is of interest to note the parallelism in time between the early development of nuclear energy and of commercial wind-electric generation in this country, as presented in table 7 on page 200.

Thus the only real choice between the options of nuclear power and wind power was made in the early 1950s, at a time of unjustified nuclear euphoria anticipating nuclear power ultimately so troublefree and cheap that it would hardly pay to meter electricity. The author,

TABLE 7. **Timetable of Early Energy Development**

Year	*Nuclear*	*Wind*
1938	Fission discovered in Germany	Putnam proposes construction to Smith Company.
1939	News of fission reaches U.S. Research and development started.	Smith Company starts construction of Vermont wind turbine.
1941	First nuclear chain reaction, Chicago.	Turbine completed, first commercial wind generation of electric power in U.S.
1945	Atomic bombs developed and used to end war. Public learns of nuclear energy.	Vermont commercial demonstration terminated, wind power found uneconomic compared with hydroelectric and cheap fuel.
1953-54	Decision to sponsor commercial nuclear power under slogan "Atoms for Peace."	Federal Power Commission proposal for construction of a much larger wind generator as a next step in wind-power development died in congressional committee.

working in a national laboratory, shared this hopeful illusion at the time. Despite all the disappointments with nuclear technology in the meantime and a growing appreciation of its dangerous international implications, we are still operating with the choice of options made under the influence of that euphoria. In the intervening quarter century, nuclear power has been given enormous public sums covering activities all the way from uranium prospecting and fuel preparation through development studies and preferential liability and insurance provisions to waste disposal, and has also been permitted to draw heavily on increased electric rates; while our government paid no attention at all to wind power until the seventies.

The government's recent look into this and other solar-related sources began in the National Science Foundation (NSF), whose mission is development of ideas rather than hardware. It was expanded to include some demonstration of hardware in the Atomic Energy Commission where, not being atomic, it received scant support. Congress became concerned about the inequity of the active promotion of nuclear power by the Atomic Energy Commission, which was set up largely for that purpose, and the passive academic interest in solar power. In order to improve this situation, and incidentally to assign the regulatory function to another agency, Congress formally abolished the AEC and set up ERDA, the Energy

Research and Development Authority, in its place. The intent was clearly to place the development of all prospective energy sources on an equal footing in the same promotional agency.

There was the practical difficulty that almost all the people in government working in energy-related fields were in the AEC so the reorganization amounted to little more than a change of name, deleting the word "atomic" to make room for the transfer of solar and other energies into the new organization from NSF and elsewhere. Solar energy obviously did not enter as an equal partner. The commitment of personnel interest remained preponderantly on the nuclear side.

During the oil embargo of 1973 the president asked Dixy Lee Ray, then head of the AEC, to prepare recommendations for a national energy development program. A number of prestigious study panels were set up and carried out studies and made their reports to be synthesized in a final report and policy recommendation. There was a Panel IX on solar power. Its report concluded that solar power, including wind power, could make a substantial contribution to our needs and recommended that it should be promoted by a reasonably accelerated program that it specified, or at least by what it called a "minimum viable" program. The final report to the president (Ray, 1973) included the reports of all the other panels but made no mention of Panel IX, whose report was suppressed until forced into the open later by litigation under the Freedom of Information Act.

The final report referred only vaguely to the possibility that renewable energy sources might make some contribution in the next century with no supporting evidence for this pessimistic view and no mention of there being more hopeful evidence in the relevant panel report. It proposed funding at less than half the level Panel IX had called the "minimum viable" (Welch, 1977). Thus wind power was officially decreed to be something for the next century. It was simply assigned a lowly place by executive fiat. Since this dictum has been influential in the allocation of development funds, it has so far been a self-fulfilling prophesy. It remains intact as official policy. It is stated explicitly in the summary of ERDA's *National Plan for Energy Research, Development and Demonstration* (ERDA, 1976) which, in fourteen pages, mentions wind power only once where it describes national goals thus: "Long Term (Beyond 2000): Permit the use of essentially inexhaustible resources" including "solar electric energy from a variety of options including wind power. . . . " Note the presumptuous word "permit," as though it were planned to keep the lid on until then.

This postponement of the hopes for solar-generated electricity required ignoring not only the suppressed Panel IX report but also a similar open report of a 1973 study by RANN, the Research Applied to National Needs program of the National Science Foundation (RANN, 1973; see also Donovan, 1972). For a program of "proof of concept experiments" it recommended an "accelerated program" demonstrating and testing large wind dynamos in substantial numbers according to the schedule in table 8.

The same report also described what it called a "minimum program" differing from this by omitting the 1978 column and postponing the figures in the two previous columns by a year. These programs, particularly the accelerated one, envisage simultaneous demonstration of several models of wind dynamos in each size category in order to provide a variety of experience as a background for mass production. As the report says, the "accelerated program provides for parallel development paths and redundant tasks, perhaps by independent investigators, to provide reduced technical risks and increased probability of success." The report adds, "Since wind energy systems, if mass produced, would appear to be very competitive with fossil and nuclear plants in the near-term, it would be expedient to start large-scale production, installation and operation of commercial systems of this type as soon as possible."

When the solar effort was taken over by the AEC and then ERDA, the development was stretched out by means of the policy directive that within each program alternative large-scale designs to reach the same goal should be demonstrated sequentially rather than simultaneously. This leads to the much smaller numbers achieved in the actual program, little more than one-tenth of those of the "minimum program," as indicated by the numbers in parentheses in table 8. These seem consistent with a policy to "permit" the massive use of large-scale wind power only after the turn of the century.

TABLE 8. **Numbers of Demonstration Wind Dynamos Proposed**

Fiscal year	1975	1976	1977	1978
100 kilowatt	4	12 (1)	40	150 (4)
1 megawatt		4	13	50 (1)
5 megawatt			4	12
Cost per kilowatt (Estimated in 1973 $)	$5,000	$2,000	$1,000	$500

N.B. The numbers in parentheses indicate the numbers of units being achieved in the actual ERDA-DOE program.

Present Wind-Power Program and Its Future

When the name was changed to ERDA and its Wind Power Conversion Branch was established in the Solar Energy Division, it was natural that its scope and funding would be within the limits determined by that declared policy and, indeed, its personnel chosen for compatibility with it. Seen in this light, the pace and philosophy of the program makes good sense although it seems frustratingly slow and cautious to those who would like to see wind power become an important national resource soon. Started under NSF in 1972 with a modest budget of one-fifth of a million dollars, its budget has been rapidly expanding but up through 1976 it has spent less than 20 million dollars on a large variety of mostly small projects and studies. That is less than 1 percent of what has been spent in various ways on nuclear power in the same period.

These relatively modest funds have been allocated quite cautiously, perhaps largely because the spending is done under the eagle eye of the Office of Management and Budget (OMB), directed by an administration dedicated to the rapid expansion of nuclear power. The OMB is the administration's implement for controlling what is actually done with moneys appropriated by Congress. The wind power branch of the ERDA probably finds it easy to obtain approval of small studies that seem to have a reasonably long-range objective while it could not get approval of a large wind dynamo construction project before it has paved the way with smaller ones, regardless of what past experience with quite large wind dynamos in various countries might suggest.

The program seems reasonably well designed to meet the limited objective of cautiously building up step by step to a convincing demonstration to attract the interest of industry, being very careful not to make a mistake that would either conspicuously waste money or give wind power a bad name on the way. Perhaps the mistake that was made early in the design of the Smith-Putnam machine and was recognized by the engineers but could not be rectified in wartime, which led to the throwing of a huge blade near the end of the project in 1945, accounts for some of the caution in the present program. With only provisional backing, the wind-power administrators understandably want to be very sure not to have any such accident at least until they have several demonstration machines working nicely to show what wind power can do. In their first substantial construction projects they are therefore doing essentially nothing that has not been done before and have been avoiding until later anything so innovative as a floating offshore wind turbine, though preliminary feasibility studies of this proposal have belatedly been initiated.

On the other hand there is funding for smaller investigations of a number of innovative concepts, both in theoretical studies and small-scale experimental demonstrations. These seem to be serving as a reasonable hedge against the possibility that one of the unconventional approaches, such as a vertical-axis turbine or a shrouded wind turbine, might turn out to have better economic prospects than the conventional horizontal-axis machine in spite of earlier indications to the contrary.

A rather substantial fraction of the available funds has been spread over a large number of paper studies concerned more with what to do with rather large numbers of large wind dynamos, if we had them, than with actually getting to the point of being ready to produce them. The siting studies for specific localities and the various wind distribution and acceptability studies may be expected to be useful eventually, and some of them serve to arouse the interest of utility companies in wind power before its availability. But if the purpose were to achieve commercial production of large wind turbines as quickly as possible with available funds these peripheral studies might seem premature, more a matter of deciding how to cook the chicken before it is hatched.

What was actually constructed under the program in its first four years was one 100-kilowatt wind turbine generator at Plum Brook, near Sandusky, Ohio, conveniently near the Lewis Center of the National Aeronautic and Space Administration (NASA) near Cleveland, which was responsible for its design and construction. Its power rating is less than a tenth that of the 1.25-megawatt Smith-Putnam machine that was designed and built in less than two years, over thirty years ago, but it is rated at 100 kilowatts in an eighteen-mile-per-hour wind, whereas the earlier machine was rated at 1.25 megawatts in a thirty-mile-per-hour wind, so it actually is somewhat more than a tenth as large in rated effectiveness.

The Plum Brook turbine is intended as a research-oriented machine and as such is far from economic, a fancy test frame on which interchange of parts might be used in the future to explore for optimum designs. It was conceived as the start of a slow program of development to perfect designs before committing funds to larger machines.

It is rather embarrassing that, in spite of the slow pace and the caution to avoid mistakes, a design mistake was made and the machine was operated only about sixty hours in its first year, during which it was found that the blades were subjected to greater stresses than had been calculated. The cause of the trouble was the wind shadow of the tower, which has been a well-known problem for all wind turbines with the propeller on the downwind side. The Smith-Putnam design provided for these stresses by supporting each blade on a hinge permitting independent coning freedom, a complication that may be avoided in newer machines either by sturdy enough design, by streamlining the tower, or else by having the propeller on the upwind side as was done, for example, in the successful 200-kilowatt machine at Gedser, Denmark. The correction of the mistake at Plum Brook has been to remove the stairway from the tower, substituting a cable hoist. That the designers of that elaborate machine, constructed with little thought of economy, should have done something so crude as to put a stairway in the tower cannot but suggest that they were almost unaware of the tower-shadow problem. This raises the serious question of whether it is wise to put the fate of wind power development in the hands of an agency whose personnel, while very capable in their field, have had more experience and interest in an activity so very different from wind power as space exploration—where the whole philosophy must be to use extreme care and spare no expense to avoid mistakes in a very complex technology. It is clearly an effective way to find employment for an agency whose primary mission is being curtailed and it provides ERDA-DOE with an opportunity to make use of an organization already assembled and ready to take on a job, but it may be hard on the prospects of wind power.

The next step in the DOE-NASA wind-turbine construction program consisted of contracting for two identical 1.5-megawatt machines. These were to have fiberglass blades with a span of 200 feet, and towers 150 feet high, in each respect slightly larger than the Smith-Putnam machine. Their original rating of 1.5 megawatts was for a twenty-two-mile-per-hour wind. The contract for the design and construction of this machine was let in 1976, the year following the completion of the Plum Brook machine. The contract specifically called for design and construction of the first machine at $7.5 million with an option to build the second for an additional $2.5 million, thus most of the cost of the first one being for development. The second machine was later canceled, some changes being needed.

The first machine, its rating stepped up to 2 megawatts for strong winds, is sited on Howard's Knob in the northwest corner of North Carolina, with completion expected in late 1978. The DOE is closely following the Grandpa's Knob precedent thirty-seven years later with a larger two-bladed wind dynamo on a higher windy Appalachian "knob."

As for floating offshore wind turbines, their detailed design and construction will perhaps come along later. Only a preliminary feasibility study has been undertaken as of 1977, and this only after prodding by Congress. ERDA had apparently planned to delay this investigation until more machines were operating on land, but the fiscal 1976 appropriation bill contained an amendment initiated by Senator Kennedy, instructing ERDA (as soon as practical and consistent with feasibility studies) to include in a future budget request a proposal for the construction of at least one floating offshore unit in the 2-megawatt range. This is an interesting example of Congress, or at least some of its members, being ahead of ERDA on wind power. The original wording of the amendment was more emphatic, instructing ERDA to go ahead with the construction as soon as feasible, but in Senate-House conference it was watered down, under the initiative of Congressman Mike McCormack of Washington (who formerly was an engineer at the Hanford works of the Atomic Energy Commission), to move along more slowly. The same appropriation bill contained another amendment, initiated by Senator Humphrey, directing ERDA to have at least 25 megawatts of wind power functioning by 1982 and thus applying pressure for quantity rather than diversity of experience.

The 1977 reorganization of the government's energy agencies shifted the wind-power program from the former ERDA to the new Department of Energy, DOE, apparently without changing the nature of the program very much. The intentions for the future of the wind energy program are indicated graphically in figure 47, which is more specific than the accompanying text of the 1976 energy plan (ERDA, 1976). This plan was announced almost a year after Congress passed the Humphrey amendment calling for 25 megawatts by 1982 and the next-to-last horizontal line, Multi-Unit Demonstrations, has been updated to meet this requirement and indeed exceed it by a generous margin in calling for "100-Mw Demonstrations" by the end of 1981. The use of the plural denotes presumably more than one such demonstration and thus the equivalent of about 150 or more 1.5-megawatt windmills. This, following only about a year after the first 10-megawatt demonstration of "wind power farm," represents a drastic change of pace and suggests that the Wind Energy Conversion Branch of ERDA-DOE has recognized the desirability of a rapid expansion of their program now that Congress has expressed some enthusiasm for wind power.

There is a strange disparity between this suggested goal of demonstrating a hundred or more large wind dynamos promptly and the subsequent long delay before the next forward step expressed in the last line of figure 47. This indicates that factories for mass production of wind dynamos are not to be achieved until at least twenty years after that sudden spurt of activity, sometime on beyond the thin black column labeled "1986–2000" and on into the next century. A plausible explanation for the disparity could be that the wind-power planners found it expedient at this point to defer to the policy that wind and other solar-related sources shall not be expected to make really substantial contributions until the next century.

If the inertia of decision making could be overcome and large-scale wind power could be given the priority it merits in the present national energy situation, it would be more reasonable to modify the schedule of figure 47 about as shown in figure 48, aiming at accumulating broad experience fast and entering the mass-production phase in about 1981.

The 1976 ERDA plan, if figure 47 is to be taken seriously, thus seems to include at least meeting the goal of deploying 25 megawatts of wind-electric generating capacity by 1982, as set by Congress, and probably exceeding it. There is a considerable flexibility in the details of what sort of experience this deployment is to bring. It could mean building many almost identical machines in the 1- to 2-megawatt range, perhaps similar to the one now being designed. Or it could mean testing out a dozen different designs to find out which are best and provide a base of broad experience in preparation for mass production. The trend of DOE planning now seems to be directed toward emphasizing experience gained with the use of wind dynamos "in a user environment" rather than variety of wind-dynamo designs, but that could change. Seeking a "user environment" means cooperating with various electric utility companies in the financing, construction, and tying into the power grid, using windmills of a proven design, and thus implies constructing a number of rather similar, if not quite identical, machines. The text of the plan (ERDA, 1976, vol. 1, p. 67) expresses reasonable intentions without being specific on this point:

In wind energy conversion, the program strategy is to stimulate industrial efforts to design more efficient rotor systems and to lower capital costs through prefabrication and more efficient production techniques, and through demonstrations of reliable, economically viable wind energy systems.

Such a program seems to be designed to have the government carry wind-power development and deployment only as far as is needed to interest industry in carrying it further. It may thus have

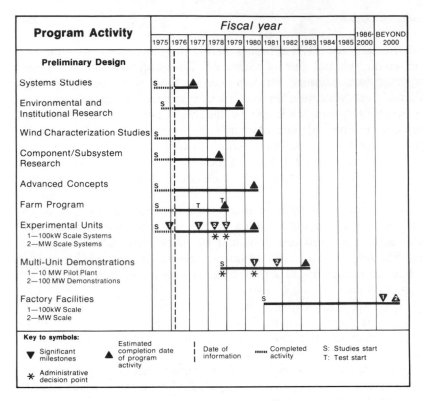

Fig. 47. Plan for federal wind power development and demonstration from a budget proposal in 1976.

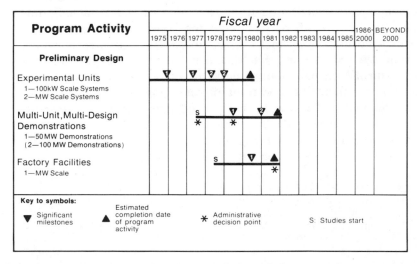

Fig. 48. Proposed accelerated schedule for wind energy demonstrations and production.

the advantage of minimizing government participation in a development that should belong to the private sector but needs a push to get started, now that the government has subsidized competing technologies. However, the success of the program depends on having selected the right design in the first place and on its being economic even when the machines are being built in rather small numbers.

The pace of the subsequent expansion would depend on the rate at which other utility companies become impressed by profits made by those who have installed some of the early machines. The emphasis is apt to be on individual machines or small groups of them in the local territories of the various utility companies rather than on very large systems centered in the windiest regions available nationally. It can be expected that the expansion under these circumstances would not be very rapid.

Recent Experience Abroad

While the 100-kilowatt and megawatt-scale part of the United States federal program is confined to a single style of wind dynamo with two blades downwind from the tower, some breadth of perspective may be gained from the current and near future demonstrations of other designs in other countries. This will not be a completely satisfactory substitute for broader experience in this country, for current engineering experience close at hand seems to be a more practical guide than that long ago or in other industrial environments.

In Denmark, besides the large independent machine at Tvind cited in appendix 4 (see fig. 59), there is a two-year national program testing three different wind dynamos to explore the effectiveness of three-bladed upwind rotors. The old Gedser mill of figure 8, after a decade of idleness, is being used and tested anew. Two 600-kilowatt machines are being built, each with three blades 40 meters in diameter on a 45-meter concrete tower. One has stayed blades, following the Gedser precedent, but with some limited pitch variability. The other has normal cantilevered blades with full pitch control (JBF, 1977b).

In the Netherlands, where the land area is considered limited, though sufficient along the coast for supplying about 20 percent of the nation's electric power from the wind, a study is being made of the prospects of offshore wind power to exploit the North Sea winds. The Dutch once used wind power to extend their land into the sea. Now they may go further to sea for their wind power. It is to be hoped that they will, particularly since we seem to be taking no such offshore initiative. A Darrieus rotor 5 meters in diameter with a free-standing shaft was tested in 1977. It is to be followed by a Darrieus turbine and a propeller-type turbine, each 25 meters in diameter for purposes of comparison.

The Canadian wind-power program, too, has concentrated on the Darrieus rotor. The world's largest, on the Magdalen Islands in the Gulf of St. Lawrence, is 80 feet in diameter and rated at 230 kilowatts in a thirty-mile-per-hour wind.

In West Germany and Sweden, as in this country, the programs have evolved from the successful Hütter experience represented by figure 10 and are confined to two-bladed downwind rotors. The recent Swedish program started with a 75-kilowatt machine with a concrete tower, somewhat smaller than the similar Plum Brook one in the United States. The German program has two parts. One is aimed mainly at export to remote islands; completion of a 200-kilowatt model being expected in 1978. The other, in the megawatt range for industrial power, includes plans for a 2- to 3-megawatt machine by about 1980.

These federal programs, like ours, are conservative and of modest size relative to the need. Decisions for more forthright programs everywhere seem to be difficult, particularly in the light of prior nuclear commitments.

The Big Wind-Power Option

The forthright way to turn us on a path toward making full use of wind power on a large scale soon would start with a government decision to take a vigorous initiative and supply the necessary funds to accomplish this goal. An important early objective should be to exploit the economies of mass production as soon as this can be based on adequate experience and dependable design. Past experience is sufficient to make it quite likely that a design close to the one being demonstrated in the 2-megawatt wind dynamo now under construction will be satisfactory for mass production and one possibility would be to begin production runs on the basis of whatever design improvements may be suggested during tests of this machine.

However, mass production could proceed on a firmer footing if it could be based on broader experience with quite a number of machines exploring the effectiveness of several variations in design details. An intensive exploratory broad-experience program of this sort would take a couple of years or so. It would not be necessary to delay the beginning of quantity production until after then: a preliminary production run of the model could proceed simultaneously, and perhaps be associated with a "100-megawatt demonstration" already tentatively included in the ERDA-DOE plans in figure 47.

It may be difficult to obtain an all-out commitment to go the whole way with wind power on the basis of the sporadic past experience, but it should not be so hard to get agreement to mount a large and varied demonstration and test program for the purpose of validating the wind-power option. A further decision can be made to exploit the option fully once it is convincingly established. The expenditure required for such a broad-experience program is very small compared with that being made to "validate the nuclear option," as the nuclear development has been described (Ray, 1973). It should be, however, perhaps ten times as costly as the present demonstration with its two similar megawatt-scale wind dynamos. The cost at this stage need not be a great obstacle, since it need not be any greater than for the 100-megawatt demonstration already contemplated, but the change in philosophy—the introduction of diversity to foster the option to start mass production very soon —would require at least a tentative decision to challenge the established policy of leaving important contributions of wind power until the next century.

The next stage, exploiting the option and providing the economic stimulus to achieve mass production quickly, would require a forthright policy decision and probably considerably increased expense. Since there is no vested interest to mount the kind of lobbying that successfully promotes some large government expenditures, this decision would have to come as a result of enlightened leadership either in Congress or, perhaps more likely, as a vital part of the energy planning of the administration. Pressure from citizens' groups might be helpful especially to get Congress to go along with the president on this.

Carrying out such a broad-experience program in two years or so would require some reorganization of the contractual procedures of the ERDA-DOE Wind Energy Conversion Branch unless the large wind power demonstrations can be passed on to the Resource Applications Division in the DOE, as is discussed further in appendix 4. The functioning of the Wind Energy Conversion Branch has developed smoothly, little affected by the transition of the top management from the Atomic Energy Commission (AEC), to the Energy Research and Development Agency (ERDA), to the Department of Energy (DOE). These changes in acronyms were motivated by the need to unify energy planning and also to give alternative energy sources an equal opportunity for development. They did little to change relative emphases and left the nuclear effort dominant. The DOE has quite a different organization chart from its predecessors but solar-related and nuclear energy development are still grouped together, this time even under the same assistant secretary. This grouping may make it still difficult to achieve a decision to pursue wind power in a big way.

If, however, a decision can be made that the Wind Energy Conversion Branch should expand its megawatt-scale demonstrations and carry out a broad-experience program quickly, this office should be strengthened in such a way as to change its outlook, its goals, and its contractual methods and capabilities, if possible even freeing it from some of the constraints of government procurement practices. In the interest of a prompt start, the time for defining and making contracts could be considerably compressed. Rather than to require, as now, the request for proposals for design, then the design contracts, then the administrative review of the design, then the request for proposals for construction, the judging of these proposals and finally the contract for construction, the procedure could be compressed to include design and construction in one contract, trusting the contractor to be just as competent as the agency to select the best design for the purpose. This process could incidentally introduce a healthy spirit of competition.

In this connection it is interesting to recall that in Denmark the decision to rely on wind power during World War II was made on the night after the German invasion, and in a few months several wind-electric generators on a 100-kilowatt scale were built and functioning in spite of wartime shortages. In the Smith-Putnam effort, also in the face of shortages, the time from the company's first interest in wind power to generating electricity on line was seventeen months.

In the much larger peacetime effort we are now proposing, it would seem that the decision concerning design types and the details involved in letting contracts for an initial accelerated demonstration and test program could be made in a very few months. Much of the present program should be retained, some of it on a larger scale, particularly the development of innovative designs as a rather small part of the new expanded program. Some of its results, such as wind surveys locating favorable sites, would be useful soon for the expanded program.

By the time the expanded program could get underway the present program will have demonstrated one design on a megawatt scale, probably with two machines, but this will not be as broad a basis as would be desirable for starting really large-scale mass production. An expanded program should begin broadening that base of experience as rapidly as possible by negotiating for several design-and-construction contracts, each contract to be for a specific design concept, perhaps with some duplication to promote competition, and each contract to involve construction of several nearly identical wind turbines. Such duplication may seem extravagant in the short term in comparison with the present cautious program but would be a wise investment because the larger expenditures on the mass production to follow would be used more effectively, on the basis of experience with a wider variety of well-tested designs.

The variety of designs to be covered by this intensive broad-experience period should include mostly engineering modifications of the turbines and their mountings within the category of the propeller-type horizontal-axis machines, which have proved most effective. Variations could include the usual cantilevered two-bladed as well as three-bladed propellers downwind from the tower; guyed three-bladed propellers upwind from the tower; the British rotatable tripod; the leaning tower concept; variations in speed control; land-based and floating offshore mounting; high and low single-turbine and perhaps multiple-turbine towers; as well as perhaps one or two vertical-axis machines on a megawatt scale.

It may be argued that careful engineering studies on paper can eliminate all the less promising design concepts and select the best one without such varied experimentation. Such studies, while important as guides and essential to successful construction, are partly based on assumptions and cannot be considered completely reliable in selecting one good design over another, particularly in the light of the difficulty of predicting vibration troubles accurately. In this connection it should be remembered that the carefully designed Plum Brook machine was inoperative because of unanticipated vibrations. Thus a selection of one or more models for mass production could be made with greater confidence if it were based on actual full-scale tests, rather than just detailed studies. Relative cost estimates, as well as performance judgments, would be more reliable after such experience.

It is important, too, that the various models tested should at least include some designed for the windiest regions in the western Great Plains and offshore to prepare for exploiting wind power on a massive scale. A serious criticism of the present DOE program of sequential tests of similar models differing in size has been that they have been aimed at adoption in individual user environments having only moderate winds, even the prospective 100-meter diameter, 2.5-megawatt wind dynamo being proposed for a twelve-mile-per-hour average wind. This emphasis seems to be in keeping with present policy that neglects the big-wind-power option.

The more recent decision to upgrade the first big wind dynamo from 1.5 to 2 megawatts for a windy site suggests an encouraging change of emphasis.

Unless the present smaller-scale work on innovative designs, such as vertical-axis wind turbines, shows impressive success in the meantime, their continued investigation would be motivated by the possibility that they might contribute improvements in later generations of wind turbines. It would not be sensible to delay the start of mass production because of this possibility. It may be expected that there will be later improvements as the technology develops in use but not such drastic improvements as to render obsolete the first mass-produced machines. Model-T Fords were used as long as they would run while the factories built to produce them went on to make later models.

This initial experience could be confined to about two or three years and involve the construction of about fifty wind turbines of five or ten different designs. The numbers might be limited by the industrial production capacity and engineering design capabilities available for this work on rather short notice. The industrial organization involved would be a useful preliminary to the later mass-production phase, encouraging some companies to plan for substantial expansion as they gain experience.

In addition to information on the performance of individual machines of various types, this stage provides the first opportunity to learn from experience about the interaction of windmills in an array. A dozen or more of these machines, not necessarily of the same type, should be placed in a checkerboard array in a windy region such as the Texas panhandle to find out how much the turbines shield one another from the wind. There might better be two such arrays, one with a spacing of perhaps ten times the tower height and another five times, to learn how closely they can be spaced without undue interference. There would be wind-velocity monitors within the arrays, and this siting pattern would not impede learning about the performance of the individual machines at the same time. This is an area in which theory is weak and experimental information is needed for the next stage (Templin, 1974; Newman, 1977).

The cost of such a broad-experience stage may be estimated by using the figures for the contract originally let by ERDA for design and construction of one or two machines of 1.5 megawatts. That contract called for $7.5 million for the first machine, including the engineering costs, and $2.5 million for the optional second machine, presumably representing construction costs only. At this rate, counting on $2.5 million for each additional machine, the cost of a contract for design and construction of ten identical machines of this size would be $30 million. This estimate is probably too high because it takes no credit for the economies of contracting for and making several identical machines at once. A broad-experience stage consisting of five such contracts for a total of fifty machines would thus cost no more than $150 million. Ten contracts for five machines each, again fifty machines but requiring more engineering effort, would at this rate come to $175 million. Some machines might be larger, and some would be floating at sea, probably increasing the cost, but this seemingly large initial experience stage lasting two years need cost no more than $200 million. This sum is about 5 percent of government expenses in two years to develop and sustain nuclear power, as judged by ERDA's fiscal 1977 budget of $1.87 billion for nonmilitary nuclear activities: fission ($680 million including $570 million for the breeder), fusion ($330 million), nuclear fuel cycle ($150 million), and uranium enrichment ($700 million).

There is a wide gap between this kind of cost estimate, based on what ERDA recently spent in a contract to induce a company to build one additional demonstration unit, and what it should cost for the construction of one of a large number of machines on a strictly commercial basis for the sake of the power they produce. This high cost estimate of $2.5 million for a 1.5-megawatt wind-electric generator (about $1,700 per installed kilowatt, or $1,250 at the revised high-wind rating of 2 megawatts) must include a considerable sum as incentive for threading through the intricacies of government contracting methods as well as for actually building the second machine. It is, indeed, much higher than the estimates in chapter 9 of quantity production costs and is only intended to suggest a top limit of what it might cost the government to arrange a larger demonstration and testing program for the sake of broad initial experience.

The need to mount this intensive broad-experience program very soon stands as the first challenge of wind power to decision makers. It is a chance to open up the larger option of massive deployment of wind power soon and should be taken now regardless of whether it appears likely that the larger option will soon be exploited. It represents a reasonable expenditure to quickly make up for our past neglect of wind power, to validate an important option, and to maintain a flexibility of response to the needs of the future.

Stimuli for Mass Production
Chapter 4 has discussed the windy regions where vast arrays of many thousands of large mass-produced wind dynamos could be profitably deployed and the magnitude of industrial effort required

to produce them. It is well within the capability of American industry and is potentially profitable once mass production gets started, but it does require a drastic reorientation of investor interest. A development and demonstration program, such as the DOE's present one but intensified and broadened as outlined above, is one important preliminary to attract investor interest. However, if there is to be a social decision to implement this technology "far more rapidly than economic decision making would otherwise warrant" it should be assumed that further financial stimulus may be needed to induce a fast enough growth rate in so new a field of investment. As has been mentioned, other energy sources have received economic stimuli in the form of tax allowances and various forms of indirect and direct subsidy. These should also be applied to wind power. The so-called fuel depletion allowance permitting a tax write-off is proportional not to depletion but to the selling price of fuel. Wind power involves no fuel so the corresponding production allowance permitting a tax write-off could instead be proportional to the selling price of electric power generated by the wind.

Provision of guaranteed loans for the purchase of wind dynamos should be an effective inducement to a utilities industry beset with the difficulty of raising enough investment capital as the country's appetite for electric power grows.

If direct subsidy is to be used, it should probably be concentrated on the first few thousand wind dynamos, perhaps at a rate of one- or two-tenths of a million dollars per installed megawatt, in order to get through the economically difficult period of establishing mass production. It could taper off or terminate as it becomes increasingly apparent that wind power pays.

There is another very appealing way for the government to stimulate the establishment of mass production of wind dynamos. The government is already in the electric power generating business as owner of the great hydroelectric dams. As has been discussed, the greatest opportunity for using wind power as base power without any additional investment in storage facilities is to team up with hydroelectric power, saving up river flow when the wind blows hard for use as extra hydroelectric power when the wind slackens. It would be a logical and politically natural extension of the government's ownership of the hydroelectric facilities if it would supplement them by acquiring large wind-dynamo arrays. Thus the government, either directly or through an appropriate government corporation, could be the first purchaser of thousands of wind dynamos at a rate to be set with a view to getting the industry started quickly. This is perhaps the most manageable way in which wind power could contribute substantially to our electric energy needs in the next decade.

The big-wind-power option is so extensive that there is no real upper limit on its size; no apparent reason why, in spite of its diffuse deployment, it should not be capable of meeting a very large part of our future electric power needs. There will be other benign sources and it will always be desirable to have a mix of sources. How extensively the wind-power option will be exploited will depend not only on its own development but that of other sources and particularly on evolving worldwide patterns of energy use. The point is that, while one cannot say just how important it will become, wind power is potentially a very important power source. The sooner we start exploiting it in a far-ranging way, the better.

Regardless of what one may see as the ultimate goal of wind power deployment, getting started promptly on mass production after a short period of broad experience seems very important. Any substantial use of wind power to generate electricity, even if far below the ultimate potential, should provide economic savings and reduce the undesirable impact of fuel-consuming energy sources. The motivation to get started right away with mass production of large wind dynamos need not depend on appreciation of how great the ultimate potential might be. Once the economies of mass production have been achieved and incentives have induced a large shift of investor interest to this energy source, a combination of economic advantage and social concern, as well as moderation of demand through conservation, will determine how far wind power will go toward meeting future real needs for electric power. Yet it is helpful even at this early stage to recognize, as the discussion in chapter 5 suggests, that under favorable circumstances wind power is technically capable of meeting that entire need. It will doubtless never to be called upon to meet it alone but it can be a major part of a desirable mix. It is perhaps not unrealistic to think of it as getting off to a fast start and meeting a few percent of the need within a decade, then increasing rapidly in the next couple of decades, ultimately meeting as much as half of the need by the time ocean-thermal power and direct solar power will have advanced far enough to take over much of the rest, supplementing remnants of the sources on which we now depend. Even if so high a goal is considered unrealistic, getting off to a fast start toward a lesser goal would be well worthwhile.

Nuclear Options

The option of meeting a large part of our growing electric energy needs with nuclear power has been spelled out often elsewhere, but not always with full appreciation of the uncertainties involved. The option now is different from that of the early 1950s when it was adopted as government policy. While there is a strong tendency to keep the present option as a continuation of the past, a new decision and a new initiative is needed now if the growth of the program is to continue. Now is an appropriate time to reconsider the option as it now appears, comparing it with other options as they now stand.

The advantages and disadvantages of nuclear power have been discussed in chapter 7. Both are quite impressive. The advantages are mainly the compactness and availability of fuel and the fact that the power plant aspect is a going industry, if not a mature one. Overlooking for a moment the more indirect disadvantages concerned with the environment and international stability, and barring serious accident, the main disadvantages from the immediate fiscal point of view are the need for commercialization of the fuel preparation and handling with the implied uncertainties of costs, the inevitable rising cost of the limited fuel supply for reactors of the present type, and the very great uncertainty of cost and performance of the future types of reactors that are supposed to come to the rescue.

Perhaps the most influential argument for nuclear power at present is the claim advertised by some utility companies that according to their books nuclear power is saving the ratepayer money as compared with coal power. They say this despite the fact that recent rate increases have been partly caused by cost overruns of nuclear power plant construction. This claim, made only by those utility companies whose nuclear plants have been doing relatively well, does not take into account that the ratepayer is also a taxpayer. Some of his taxes pay for part of the nuclear enterprise that has been and is still carried by the government. The apparent savings have two main reasons; the first is that there have been recent increases in both coal and uranium prices and fuel costs are a larger part of the total cost of power from coal, since nuclear plants are more capital-intensive. The second is that mined coal is in artificially short supply and its price is high because of the great expectations for nuclear power that led to shutting down some mines, and the recent lapses in nuclear power that have increased the demand for coal. This is apt to be a temporary situation to be changed as the price of relatively scarce uranium increases more than that of coal, which is plentiful.

All the options for nuclear power policy to be considered now must start from the present status of nuclear power as outlined at the end of chapter 7. Among the possibilities that would depend mostly on wind and other solar-related sources supplementing fossil-fuel-fired plants and not much on nuclear power, there are two options—the "no nuclear" option and the "no more nuclear" option. The first one means ceasing to rely on nuclear power immediately, decommissioning all operating plants and compensating their owners from the public purse for their investment. This option may be ecologically desirable but it is costly and probably politically impractical to forego the power from the plants that have already been built at great expense.

The second or "no more nuclear" option would permit these plants already built, and perhaps those nearing completion, to live out their useful lives while the enriched uranium supply lasts, under strict regulation to keep them as nearly safe and ecologically benign as possible, and would stop all plans for expanding nuclear power. The so-called Zero Growth Scenario includes this "no more nuclear" option, and depends on other sources to increase power production slowly before leveling off.

A third option, which might for an interim period depend heavily on both nuclear and wind power, we might call the "no breeder" option. It would exploit the present technology of water-moderated reactors as far as economically practical, with all due precautions, but would stop expenditures for breeder reactor development, since the combination of economic and safety uncertainties about the performance of this untested device, together with the implications of the resultant plutonium economy, makes it undesirable to base our plans for future power development on it.

The fourth option is the all-out nuclear option, represented by present official policy that anticipates only a small contribution from wind and other solar sources in this century. It includes not only expenditure of public funds necessary to promote the growing exploitation of the dwindling uranium supply by present technology, just as in the third option, but also increasing the momentum of commercial interest in the breeder reactor before it has been completely developed, with confidence that it will take over when the supply of even imported good uranium ore gives out. Until 1974 this policy anticipated having 1,000 big nuclear power plants by the end of the century but by 1976 this expectation had been reduced to 450 to 800 plants, about half of them breeders.

Within each of these four broad options are various more detailed options. Within the first option, for example, emphasis may be distributed between the several solar-related sources. In the fourth, the all-out nuclear option, the present government policy is to rely most on the hopes for a particular breeder reactor, the Liquid

Metal Fast Breeder (LMFBR) that has come to be known as *the* breeder, although there are five other types of breeders or near-breeders whose development could be included in such an all-out program (Weinberg, 1976). One variation of the all-out nuclear option would be to push the development of several of these, perhaps avoiding a commitment to any of them by having it done entirely at government expense, rather than increase commercial interest by inducing industry to bear part of the expense as was done to a small extent (14 percent) in the controversial Clinch River prototype in Tennessee. This would mean that while the public ultimately foots the bill, the cost of the development would be channeled entirely through taxes rather than partly through electricity rates. It would have the advantage of letting the decision about using any of the breeder types wait until the facts are in, less influenced by the momentum of commercial investment. Avoiding commercial investment and vested interest in the one particular breeder, the loop-type LMFBR, would leave greater freedom of choice even if the breeder route should seem mandatory at a later date.

Foreign developments have followed our lead in opting for liquid metal fast breeder reactors, though they do have two types of LMFBRs, the simpler pool type as well as our technically more ambitious loop type. Several types of breeders and near-breeders will remain not thoroughly investigated as programs now stand.

In spite of the uncertainties surrounding the prototype LMFBR (which would be of the loop type), at Clinch River, Tennessee, much of recent United States optimism about breeder reactors derives from the successful first two years of operation of the multinational fast breeder near Avignon in France, the 264-megawatt pool-type Phénix (Vendryes, 1977). One of the parts of any LMFBR most demanding of perfection is the boiler where heat is transferred from liquid sodium to water. These two fluids react violently if they should come in contact through a leak. After the encouraging first two years of operation, Phénix suffered a protracted shut-down for repair of such a leak, which reduces confidence in the future of breeders that was being based on its performance.

Despite the confidence with which the all-out nuclear option has been adopted until now, there is no assurance that pursuing it will be all clear sailing. While the technical expectations that the LMFBR will generate electric power and breed plutonium are probably reliable, there can be very little confidence in estimates of performance factors, breeding ratios, or costs. Experience with water-moderated reactors has amply demonstrated that they can usefully generate electric power but has not been at all reassuring about plant factors or likely future costs. The favorable costs that have been noted for nuclear power relative to coal-fired plants have been possible not only because of government support of collateral services but also because the nuclear plants concerned were ordered and mostly paid for before the recent drastic increase in the capital costs of nuclear plants. It is not likely that similarly favorable costs will prevail for power generated by the more expensive plants ordered now or later. There is no reliable indication that the disappointingly low plant factors will not sink still lower as the end of the useful reactor life approaches, which would further increase the cost of nuclear power. The magnitude of the appropriation for the proposed rescue operation called the "Nuclear Reassurance Act," which failed to pass Congress (about a hundred billion dollars) suggests the enormous likely cost to taxpayers in the next few years if the all-out nuclear option is continued.

As compared with the "no more nuclear" option, the all-out nuclear option during the next ten years represents not an increased supply of energy but a net energy sink, an added load to existing generating facilities. Every nuclear power plant ordered now would require for its construction an amount of energy equivalent to something like one-fifth of the output of such a power plant for the ten years it is being constructed. This is another way of saying that the time required for a nuclear plant to pay back the energy required for its construction is something like two years. It is only later, after the difficult ten years, that the effort would begin to pay off in additional power, and then only if continued expansion is not too rapid (Lovins and Price, 1975).

In contrast with this, the energy pay-back time for a wind dynamo is estimated to be only seven months (Coty, 1977). The time required for its construction is about a year so it finishes paying for itself in energy and starts contributing net energy much sooner after construction is started.

From the commercial and financial point of view, ignoring social-ecological considerations, and with confidence that the breeder will be ready to provide continuity of fuel supply, the uncertain future cost of nuclear power is not a reason for abandoning the all-out nuclear option, because the regulatory structure ensures that electric rates will be increased to cover increasing costs and essentially guarantees utility companies a profit proportional to the amount of power sold. Thus the generating cost is not relevant because of its effect on profit margins but only because of its effect on selling prices influencing the amount of power sold. From this standpoint the important consideration is that reactors will provide the power with a continuity that justifies past investments and that the government can be persuaded to continue providing the collateral services. It can be expected that there will be continued strong pressure for extending the earlier commitment to the all-out nuclear option in spite of its drawbacks. Any moderating influences there may be on this pressure may come from doubts about continued exponential energy growth and inadequacy of available capital to finance both the growth of generating facilities and adequate use of the energy.

Summary and Conclusion

Wind power, the most completely developed of the solar-related sources, is technically ready for immediate promotion and a rapid expansion program that could soon supply an adequate supplement to fossil-fuel sources of electric power in a reasonably growing economy. This would lead naturally into the future situation in which it and other solar-related sources will carry most or all of the electric power without adding a massive heat burden to the biosphere. Any claim that there is no alternative to nuclear power is based on the assumption that adequate promotion will not be initiated, that wind power must grow slowly like the railroads or the automotive industry through cumulative commercial interest, although in competition with other power sources that have received heavy government subsidy and promotion. It might instead grow quickly like the transistor industry that achieved mass production through demand sparked by government purchases.

Cost estimates of wind power based on government contracting for design and construction of demonstration models, employing personnel accustomed to space-agency methods, are unrealistically high. Costs estimated on the basis of knowledge gained by building large experimental wind dynamos in various countries, updated to account for inflation and corrected for differing plant factors (including that arising from wind variability), show that dependable electricity generated on a large scale by wind power with adequate storage will probably be considerably cheaper than that generated by future nuclear power plants. Since we have a larger potential for wind power in our spacious windy regions than do more crowded industrial countries, introduction of large-scale wind power here to substitute for some or all of our projected nuclear growth would give us a competitive advantage. We could continue our tradition of favoring American industry with the availability of cheap power, compared with competitors who cannot so readily find substitutes for nuclear power. Introduction of large-scale wind power here, demonstrating its economic feasibility, would also have the great advantage of providing an alternative to exporting nuclear power plants to the less developed countries. We would compete against nuclear exports from abroad with a power source better suited to the talents and needs of those countries and reduce the extent of the dangerous proliferation of nuclear materials.

If our government were to decide to promote a rapid introduction of large-scale wind power, the logical first step would be to establish quickly a base of broad experience with an aggressive demonstration program involving construction of several dozen megawatt-scale wind dynamos of a variety of designs simultaneously, rather than sequentially as in the present slow wind-power demonstration program. This could take two or three years and would provide the basis for initiating mass production through appropriate cooperation of government and industry shortly thereafter, so that the wind could begin to contribute substantial amounts of power in less than a decade.

If the government cannot decide initially to go the whole way, it seems very important that it should decide to keep open the option to do so by initiating the broad-experience program immediately, postponing until two or three years from now the decision whether to go ahead and promote the mass-production stage. The cost of such a broad-experience program would be small compared with what is being spent on the development of other sources of power.

Promotion of the mass-production stage could involve direct subsidy or various tax advantages similar to some that have been enjoyed by competing power sources for many years. One appealing method, analogous to the way mass production was achieved quickly in the transistor industry through a ready government market, would be for the government to supplement its extensive hydroelectric power with wind power to carry the load while the wind blows, saving the limited river flow to supply more hydroelectric power when the wind slackens. Since the government is already in the business of generating base power that is sold to utility companies, this would be a natural extension of present activities and would make up for some of the underperformance of hydroelectric installations that is due to greater evaporation and reservoir leakage than was anticipated when the generating facilities were designed. After this demand has initiated mass production, suitably encouraged private investment would be expected to establish huge arrays of large wind dynamos, in the windy regions of the western plains and offshore, which would feed into extensive electric grids.

The transition of emphasis from one technology to another, from continued growth of nuclear power to rapid growth of wind power, necessitates dislocations of workers and of organizations which would naturally be resisted by those involved. Workers accustomed to one type of work will in many cases have to turn to another, sometimes after some retraining. Welders working on pressure vessels may shift to making windmill towers. The fact that the new industry would create more jobs than are lost should be convincing to planners seeking to reduce unemployment but is not impressive to the individual jobholder nor to his union. If the workers could be assured in advance that those displaced from the old industry would be given hiring preference in the new, this might ease resistance to the change and facilitate the political decision to make the shift.

This political decision has been opposed, so far successfully, by the vigorously promoted and widely accepted claim that "there is no alternative to continued growth of nuclear power." The twin claim, appearing in successive ERDA planning papers, is that solar power, including wind, is to be lumped with far-out nuclear fusion as needing extensive R&D before being ready to supply substantial amounts of power in the next century. The pages of this book should help dispel these myths.

Proponents of various energy growth patterns for the future come close to agreeing that the rate of growth should not change too suddenly, with a sharp corner in the curve or a sudden drop, else the shortfall in planned energy would be very disruptive. In the present acrimonious contention between the proponents of nuclear power and the proponents of alternatives there may be a danger that each side might actually stop the other power source and there would be such a disruptive shortfall. It is important here to remember two things about nuclear power. First, serious technical and financial difficulties have halted its rapid growth, which may not resume short of a massive federal bailout. Second, the time scale for ordering and building new nuclear power plants is about ten years, about the time it would take to make the political decision and carry out the final testing and industrial reorganization to generate comparable amounts of electric power by the wind. While either type of plant is being built, it consumes energy as an investment rather than producing it, but with the wind the construction and the payback times are shorter.

With the pressing need to diversify our energy sources, the great opportunity of large-scale wind power should not be missed simply because an earlier choice of options at a time of cheap fuels has established practices from which we find it difficult to deviate.

Appendices

Appendix 1
The Power of the Global Winds

A great, gravity-contained nuclear reactor within the sun is the source of practically all the energy that sustains life on this planet. That energy is transmitted to us by sunlight. Sunlight is so bright that, when the sun is overhead on a clear day, it is measured as one kilowatt per square meter. We can appreciate what that figure means by thinking of sunlight as much brighter than the light from an ordinary reading lamp, say one with a 100-watt bulb in it. Ten such lamps, amounting to one kilowatt, if all their light were focused on a square meter (or square yard) would make it about as bright as would sunlight. If we could harness and use the sunlight from a square meter with 100 percent efficiency it would light those ten electric bulbs. Even with a lot of inefficiency, it would light one of them. That is what one hopes to do on a grand scale, with very many square meters of collecting surface to power many light bulbs and other electric devices.

Sunlight is brighter at the top of the atmosphere, about 1.6 kilowatts per square meter. Multiplying this by the projected area of the earth shows that the total input of sunlight to the earth is about 170 billion megawatts, about twenty thousand times mankind's worldwide use of power. But harnessing even so small a fraction as five-thousandths of a percent would be a formidable undertaking and that is what the solar-energy enterprise in general and this book in particular are all about.

The solar energy of sunlight has many effects as it reaches the earth. It makes green things grow on land and thus provides us and the animals we eat with food. It nurtures the biota of the sea on which all sea life depends for food. It penetrates through the upper atmosphere almost unhindered as visible light but, when transformed to heat rays that try to escape into the cold void of space, most of it is trapped by molecular vibrations in the atmosphere. It remains as heat and keeps the temperature of the air in a range suitable for supporting life, while the ultraviolet part of the incoming rays of the sun is filtered out so as not to destroy life.

   Among the many things it does to make life possible on earth,
the sun heats the air and the oceans with varying intensity at different places, which results in winds and ocean currents. Without
winds, the earth would be intolerably hot at some places and intolerably cold at others, much more extreme than it actually is.
Fortunately, air expands and becomes lighter when heated. A
draught goes up a chimney above a fire because the heavier air
outside displaces the lighter air inside. The same imbalance in the
atmosphere causes the winds.

   However, the circulation induced between the hot places and
the cold places on earth is on such a grand scale that the winds are
much affected by the rotation of the earth. Anyone who has ever
done the experiment of trying to walk along a straight line on a rotating merry-go-round is aware that things seem to work differently
on a rotating body. There is a strange force that seems to push sideways as one walks forward. It is called a Coriolis force. It is associated with the fact that following a straight line as seen by someone on the merry-go-round means following a curved line as seen
by someone on the ground beside it (or vice versa) and moving along
a curved line implies a sideways force. The rotation of the earth
similarly makes a railroad train, for example, tend to swerve a little
toward the right and in a straight section wear out the right-hand
track a little more than the left in the northern hemisphere and
the other way around in the southern hemisphere. Near the equator, just between the two hemispheres, the surface is almost parallel
to the axis of rotation and there is practically no such effect.

   The winds aloft are so fast and unhampered that for them this
effect is very strong in most parts of the globe. If the earth were
not rotating, the air would simply rise in regions where it is warm
and descend where it is cooler and circulate directly between,
pushed horizontally by pressure differences that arise from the different densities of the warm and cool air columns. On the rotating
earth, the wind tends instead to move at right angles to the line between the regions of high and low pressure and the Coriolis force
pushes at right angles to the direction of the wind to maintain the
pressure difference. In the confusion of winds near the surface where
we experience the weather, the wind does not flow directly from a
region marked high pressure on the weather map to a low-pressure
region, but rather tends to spiral almost at right angles to that direction, traveling a lot farther than it otherwise would to get there
if it gets there at all. This situation is known as geostrophic flow.
This perhaps gives some idea of why meteorology is so complicated,
and the weatherman is so often wrong. Even with powerful math-

ematical methods, modern computers, and satellite observations, there is no complete theory of the circulation on which we can rely. We can try to understand some general aspects to help us appreciate how powerfully the circulation is driven and how mere man could little affect it with mechanical devices like windmills.

The large-scale circulation of the atmosphere arises mainly from the temperature difference between the equatorial and polar regions of the earth. This is the most important source of the power of wind. In a broad band near the equator the sun streams in more nearly vertically than elsewhere and the heating is more powerful than the cooling by re-radiation into space, giving a net heating effect. In the polar regions the sun's rays are slanted so a given amount of sunlight is spread over a larger area but since the cooling re-radiation can go straight up, there is more cooling than heating and consequently it can get very cold there. For a part of the year the extreme polar regions actually get no sunlight at all. Also, there is more direct reflection of sunlight from the polar ice and less heating by absorption in the clear, cold air that contains less water vapor than elsewhere.

The general circulation of the atmosphere is divided into three types of zones, the equatorial zone where the Coriolis force is unimportant, the temperate zone where we shall be most interested in wind power, and the polar zones where the Coriolis force is strongest and can bend the lines of flow around the poles. In the tropics the flow approximates that of a simple thermal cell, such as in figure 49a, with the air rising at the equator and descending near north and south latitude 20° where it is slightly cooler. The pressure differences that drive the winds in such a cell are indicated in figure 49a, where the surface winds are driven from "high" to "fairly high" and the winds at any particular high altitude are driven from "fairly low" to "low" pressure. There is a greater pressure difference between "high" and "low" on the cool-air side that there is on the warm side because the cool air is heavier.

In the temperate zone the situation is very different. If it were only slightly different because of a weak Coriolis force in a slowly rotating earth, we would expect the northward upper air flow in the northern hemisphere to be deflected toward the right (which means toward the east), and the southward surface winds to be deflected toward the west—but they would be retarded by ground friction so the more important effect would be to turn the upper air flow eastward. In this slow-rotation situation, the main heat transfer between tropics and polar region would involve a return flow along the surface as strong as the northward flow aloft.

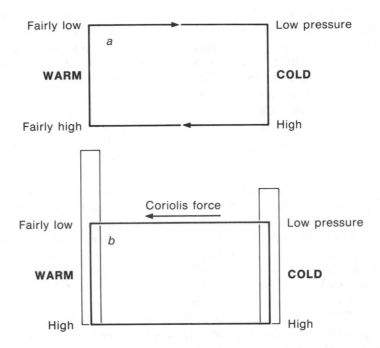

Fig. 49. (*a*) Simple Hadley-cell convection between warm and cool sides of a room, or between equatorial and subtropical regions of the atmosphere. (*b*) In the temperate zone this convection is arrested by the Coriolis force aloft due to the eastward movement of the higher air.

On the actual earth with it strong Coriolis force, this effect is overwhelmed in importance by another mode of northward heat transport by almost horizontal circulation. This takes quite different forms in the upper atmosphere and the lower atmosphere, with mutual influences by pressure but with very little exchange of air between them. The most striking feature of the motion of the upper air is the jet stream. This is a fast-moving channel of air in the lower stratosphere, about five miles up, moving sometimes as fast as 200 miles an hour, following a snakelike pattern as it wends its way between the tropics and the polar region. Sometimes it makes two round trips between these regions, sometimes even three or four as it encircles the earth, flowing eastward within a pattern, which also drifts eastward and is largely responsible for the eastward drift of weather patterns. Below it and interacting with it in a complicated way are the warm fronts and cold fronts of the near-surface winds. The jet stream drags the surface winds along, transmitting power to them that they dissipate through viscous friction especially where the air layers near the surface slide over one another with a lot of shearing motion.

While this snakelike pattern is caused by a complicated interaction of forces, it is clearly an efficient way to transport heat between a warm and cold region. The mechanics of it depend on a fairly strong Coriolis force that accompanies the earth's rotation and it can be nicely reproduced in a laboratory demonstration. A cylindrical vessel of water with a doughnut-shaped cross section is mounted on a rotating table with the center of rotation at the center of the doughnut (fig. 50). The outer boundary of the ring filled with water is kept warm, simulating the tropics, and the inner boundary is kept cold, simulating the artic region. The motion of fine particles suspended in the water is photographed from above with a camera that rotates with the table. At appropriate rotational speeds the laboratory jet stream shows up clearly, as shown in figure 51.

While the strong Coriolis force breaks up the simple thermalcell flow and imposes nearly horizontal motion, the flow in the jet stream is not quite horizontal. One of the two mechanisms driving it is gravity acting on a slightly downward and outward (or south-

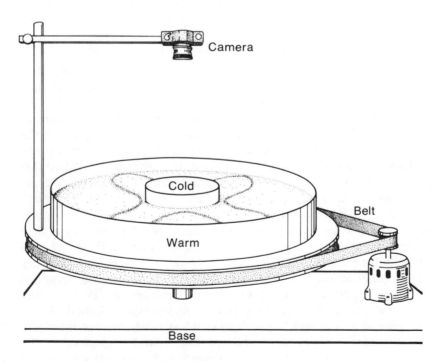

Fig. 50. Hide's annulus experiment.

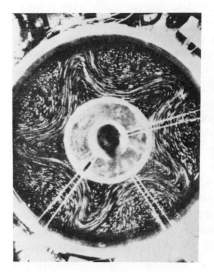 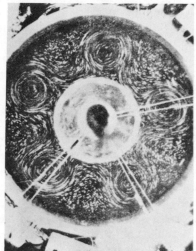

Fig. 51. The laboratory model of a jet stream on the surface of water on a rotating table. The shape of the stream varies slowly and periodically, returning to the left-hand shape after sixteen revolutions (equivalent to sixteen days) but having the right-hand shape after eight revolutions. (*Courtesy World Meteorological Organization.*)

ward) flow of cooler fluid, while there is a slightly upward flow of warmer fluid on the inward (or northward) leg of the snakelike pattern. This driving mechanism would function whether the jet stream flows toward the east or west. However, the pressure at any given altitude in the jet-stream region must be greater on the warm tropic side than on the cold arctic side, as we shall see. To be consistent with this the jet stream must flow toward the east so that its Coriolis force will push south to maintain the pressure difference. To understand this pressure difference, refer to figure 49b. Near ground level the motion is circuitous, with a lot of motion around the highs and lows of the weather pattern, resulting in only a slow average drift toward the east which is retarded by frictional coupling to the ground. Thus the average Coriolis force can sustain very little pressure difference at ground level between southern and northern regions and indeed the observed average sea-level pressure is almost the same in these regions, a pressure that is called "high" in figure 49b. The fact that the pressure at the bottom of the figure is the same on the right side of the figure, representing the arctic side, as on the left side, representing the tropical side, means that the column of air above a square meter of surface weighs the same on either side. Since the warm air on the left side is less dense, the column must extend higher to equal the weight of the column of cold air on the right, and the pressure decreases less rapidly with increasing height on the left than on the right. The whole warm column has been lifted by expansion. Therefore, at a given level, fairly high up, the pressure is greater on the left than on the right. This pressure difference is possible because of the Coriolis force arising from the eastward flow of the upper air in between. Without this eastward flow, the higher air to the left would spill over to the right and establish the cell pattern of figure 49a.

The second and probably more important mechanism driving the jet stream is closely analogous to the way gravity drives the circulation of warm air up a chimney and cooler air back down outside, but in this case gravity is replaced by Coriolis force. The general eastward motion of the air mass in the northern temperate zone causes a southward Coriolis force on each unit of volume of air, just as gravity exerts a downward force on the air around the chimney, but the force per unit volume is greater on the denser, cooler air and it flows in the direction of the force as the less dense, warmer air flows in the opposite direction. Thus the warmer northbound leg of the jet stream is driven by "floating in a sea" of Coriolis force, so to speak, while the denser southbound stream sinks into it.

There are two aspects of the eastward motion of the upper air in the temperate zone, the eastward flow within the jet stream itself and the eastward drift of the mass of air carrying along the whole snakelike pattern and the pockets of air caught in its loops. Both are important in dragging along the slower air in the lower layers in a predominantly eastward direction. Also, the very rapid flow in the jet stream is important in maintaining regions of high pressure in the loops to its right and low pressure in the loops to its left. These help to induce the highs and lows of the weather pattern far beneath. The position of the highs and lows at the lower level, however, lags behind those at the high level, being also influenced by the fact that the warmer and lighter air in the inbound leg of the jet stream tends to make a low pressure area beneath it and the colder and heavier air in the outbound leg tends to create a high. This is indicated in figure 52. The Coriolis force is again important at the lower level keeping the motions there consistent with the pressure differences between these highs and lows. Thus we see that the whole system is geared together by the pressure differences and the Coriolis force, almost like a giant clockwork but with more flexibility.

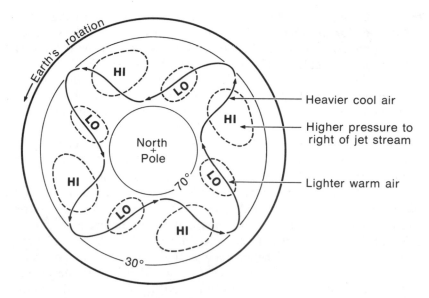

Fig. 52. The black line represents the jet stream, with its meandering exaggerated. The loops indicate high- and low-pressure regions in the lower atmosphere beneath it.

To recapitulate, the circulation in the temperate zone is driven by solar radiation that preferentially heats the air on the tropical side and keeps it warmer in spite of the way the jet stream carries heat poleward, deriving its driving force from the warm air in the process. The warmth of the air maintains higher pressure at any certain height on the warm side that drives the jet stream and associated high-level circulation. The eastward motion of the layer containing the jet stream, including the flow within the stream itself, induces a Coriolis force that holds up this higher pressure on the warm side, just as a dam restrains water, but the jet stream acts as a leak in the dam that permits a poleward flow of expanded warm air. The stream of warm air responds to the pressure almost the way pressure applied to a tube squeezes out toothpaste, and is compensated by a contrary flow of cold air propelled by the Coriolis force pushing more effectively on the higher density of the cold air. Thus there is no net flow of air between the two sides and the system remains steady, with an influx of solar energy that is dissipated by friction within the system. Power is transmitted from the layer containing the jet stream to the lower levels of the system, where we experience the weather, by friction and some exchange of air and by the pressure differences that couple the highs and lows of the two levels. Perhaps most of the dissipation takes place in the lower levels where there is shear near the ground and here there is power to spare for driving windmills. Besides the main driving force derived from the global circulation, there are other smaller driving forces that cause local complications in surface winds. Some winds are caused locally by temperature differences between places not very far apart. At the seashore the wind is often caused mainly by the temperature difference between nearby sea and land, which reverses between day and night. There are seasonal winds caused by differences between mountains and nearby plains, such as the *foehn* or *mistral* of Europe with its Alps.

The complexities of meteorology cannot be adequately treated in this brief sketch, but to go slightly into more technical detail we may note how the importance of the circulation in the upper atmosphere is portrayed in figure 53 in terms of the northward transport of angular momentum. If two streams of air cross a parallel of lati-

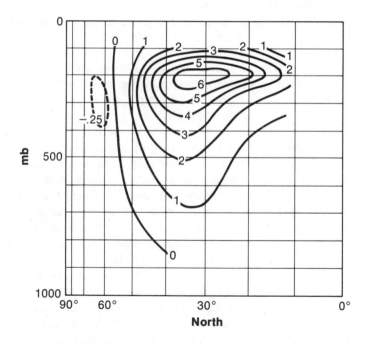

Fig. 53. Contours of equal density of northward transport of angular momentum. Altitude is indicted by pressure in millibars, with sea-level pressure at the bottom. (*Courtesy World Meteorological Organization.*)

tude carrying the same amount of air across it, one toward the north side and the other to the south side, the definition of angular momentum tells us that they carry different amounts of angular momentum unless they make the same angle with the parallel of latitude, that is, unless they are sloped the same amount north or south, one going directly northeast and the other southeast, for example. Much of the net northward flow of angular momentum indicated in figure 53 corresponds to the way the jet stream in its snakelike route tends to point eastward more on its northward leg than on its southward leg.

The northern cap of the atmosphere, above a given latitude, has a constant angular momentum. Otherwise it would be continually speeding up or slowing down. The northward flow of angular momentum into it in the upper atmosphere is compensated by the opposite torque of the westward force exerted by ground friction on the surface winds moving eastward. With this in mind, we see from figure 53 that the main driving force of windmills comes from the circulation in the upper atmosphere.

However, this high altitude circulation actually carries considerably less heat than do the winds further down, as shown in figure 54. The sharp contrast between figures 53 and 54 is largely due to the fact that lower down the temperature gradients are greater because of heat introduced into the smaller-scale eddies by interaction with the surface. For example, there is more evaporation on the south side of an eddy surrounding a typical high-pressure area a few hundred miles in diameter than there is on the north side but the moisture is retained as vapor until it circulates to the cooler north side where it condenses, incidentally releasing some heat that is being transported northward in this circulation by means of water vapor transport. The condensation makes clouds on the north side that reflect sunlight back into space and accentuate the natural tendency for the north side to be cooler. What is plotted in figure 54 is the transport of "sensible" heat arising from the temperature difference between the south and north sides. In the upper atmosphere there is no such direct interaction with the ground and less water vapor to absorb sunlight in the air so the temperature differences tend to be less and are spread out over much greater distances. Yet there is some temperature difference between northbound and southbound currents and they are propelled systematically by a larger pressure difference that is maintained, as we have seen, by the temperature differences at lower altitudes that keep the atmosphere lifted higher near the equator than near the poles. Thus, even though powered partly by heat absorbed at lower altitudes in

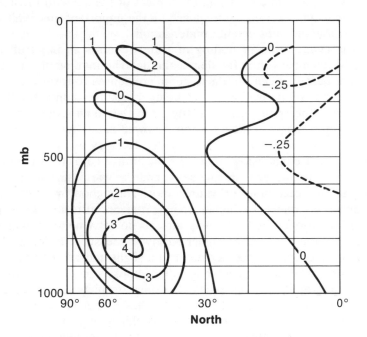

Fig. 54. Contours of equal density of northward transport of "sensible" heat, i.e., as measured by a thermometer. (*Courtesy World Meteorological Organization.*)

the tropics, the high-altitude flow accounts for a large part of the power of atmospheric circulation.

Thus the overall pattern of the weather and its variation is determined very largely by the behavior of the upper atmosphere as it relentlessly carries the heat of the sun from the tropics to the polar region. The mechanical interaction with the ground merely provides a lower boundary layer of relatively slow-moving air. If we were to erect a lot of big windmills extracting power from the wind, it would merely increase the effectiveness of the boundary-layer drag and would be expected to have little if any effect on the higher flow patterns that determine the progression of the weather, as is discussed in chapter 10. This is a complicated matter in need of much further study, but this distinction between the high-level circulation where the global patterns are formed and the slower low-level flow where power might be extracted must be an important part of the overall picture.

In addition to heating the atmosphere by absorption of sunshine in the air and clouds and by contact of the air with the heated earth, the sun heats the atmosphere by the important mechanism of evaporation and subsequent condensation.

The evaporation or boiling of water requires an input of heat, as in the example of the familiar teakettle. The inverse of this is the fact that heat is given off when steam condenses to form water. An illustration is found in the pain of being scalded by contact with live steam from the spout of a teakettle or in the condenser at the low-temperature end of a steam-turbine power plant, where a flow of cooling water or other fluid is provided to carry away the heat released by the condensing steam. In the atmosphere, the heat of the sun evaporates water from the seas, and the warm air near the surface can hold quite a lot of water vapor. Throughout most of the atmosphere, as one goes higher, where the pressure is less, it gets colder, corresponding to the way air gets colder when it expands as it moves from a region of higher to lower pressure. If some of the moist air near the surface is caught in a slight updraft, it gets colder as it rises. When air is colder it cannot hold as much moisture and some of the moisture condenses into small droplets that remain suspended in the air and the heat warms the air that surrounds them. The air may actually continue to get cooler as it continues to rise, but is still warmer than it would be without the condensation and thus is warmer than the surrounding stationary air at the same altitude. The warmer air is lighter so it is pushed upward by the weight of the cooler air around it, adding to the strength of the updraft. This may happen over a rather wide area, with the condensed droplets merely forming clouds, or it may become concentrated in the violence of a thunderstorm as the droplets churn about in the updraft and grow large enough to fall as rain. As the air warmed by condensation rises, the air drawn in along the surface is one of the forms of wind that may drive a sailing ship or power a windmill. Thus, in these various ways, solar radiation is the source of wind power.

There is so much energy stored up in the global winds that, if it could be converted directly, it would supply all of the energy used by mankind for about five years. Yet it is a renewed resource: it is driven by the energy brought in by sunlight with a power that is very great compared to any that mankind might use. This is of course an important fact for the prospects of wind power. To appreciate its extent we shall consider a few numbers.

Measurements of wind velocities at various altitudes and at various parts of the earth lead to the estimate that the total energy of the motion of wind, that is, its kinetic energy, is 330 billion megawatt-hours (3.3 10^{11} Mwh, a megawatt being a thousand kilowatts). To get some idea of what this large number means, this corresponds to an average value of the wind speed throughout the atmosphere (more technically, a root-mean-square value) of 50 miles per hour.

For those who like to calculate, this is a straightforward calculation. The atmosphere weighs enough to hold up a column of water about 10 meters high (which is as high as one can pump water with a suction pump, equivalent to the 76 cm column of mercury in a barometer). Since a cubic meter of water weighs a ton, this means that there are 10 tons of atmosphere for each square meter of the earth. The area of the earth is 4π (6 10^6 meters)^2 and the mass of the atmosphere thus 4.5 10^{15} tons or M = 4.5 10^{18} kilograms. The kinetic energy is 3.3 10^{14} kwh times 3,600 sec/hr which equals 1.2 10^{21} watt-seconds or joules or kilogram (m/sec)^2. Setting this equal to 1/2 MV^2, the expression for kinetic energy if the whole atmosphere had the same speed to get an average value, one obtains V = 23 meters/second = 50 miles per hour.

If we judge by winds we know at the surface this seems like a very high value, almost a strong gale, for the winds driving wind dynamos in selected windy regions are typically only about 20 or 30 miles per hour. There are parts of the earth where the surface winds seldom get that high. That the average is so much greater than typical surface winds is an indication of how much stronger the winds are aloft. Commercial aircraft commonly encounter winds of 50 to 100 miles per hour and in the jet stream the speed is often 200 miles per hour.

The sun's radiation falls on the atmosphere of the earth as short-wavelength visible light rays characteristic of the high temperature of the sun's surface and about a third of it is reflected right back into space, mostly from the atmosphere but about 6 percent from the surface, as short-wave light rays. The rest is converted ultimately into heat and radiated back into space as longer-wave heat radiation, the outgoing total being the same as the incoming according to the conservation of energy. The quality of the radiation is degraded from the high-temperature type to the low-temperature type by several processes, and the driving of the winds is included as a sort of heat engine, providing motion by absorbing energy at a high temperature and emitting it at a lower temperature. The energy of the motion is ultimately dissipated in some form of friction that makes heat that is radiated into space, all in accord with some physical principles known as thermodynamics.

Thus all of the energy that is put into the wind by a driving force, aside from such minor actions as driving wind turbines and breaking trees, is dissipated by friction, also known as viscosity, within the wind itself as different parts having varying speeds rub past one another, so to speak. This rubbing of shear is especially pronounced just above the surface of the earth, between the lower layers that are slow or almost stationary because of friction with the earth and forests and such, and the stronger winds higher up. A substantial part of the dissipation of energy takes place there. Analysis of observations of wind speeds throughout the first few hundred meters of height in various parts of the world leads to the estimate that the global dissipation in this layer near the surface is about three billion megawatts, 3 10^{15} watts (Hütter, 1975). (German and British sources give the figure 2.5, Russian 4.2.) This is about 2 percent of the incoming solar radiation. This seems to be as good an indication as can be obtained of how much the sunlight's energy is effective in driving the winds, yet this is only the part of it in about the lower tenth of the atmosphere. While there is no boundary like the earth's surface to induce such a rapid variation of speed higher up, there must also be a great deal of shear in the churning motions and near the edge of the very fast jet stream up there. It would seem surprising if the total were not something like twice this value or more, perhaps 4 percent of incoming solar radiation. The amount of the sun's energy driving the winds (and waves) is commonly quoted as about eight times less than this, 3.7 10^{14} watts (Hubbert, 1973 and several later textbooks), but in that accounting evaporation is listed as a separate item. A small part of the driving force of the winds must come from the small percentage of sunlight that is absorbed in the air and heats it directly, but the larger percentage of sunlight that causes evaporation must have a larger role in driving wind.

Of the sunlight that reaches the earth's surface, which is about two-thirds of the total, most is directly converted to heat and re-radiated, with quite a lot of bouncing back and forth between clouds and earth on the way. About 23 percent of the total, 40 billion megawatts or 4×10^{16} watts, is spent evaporating water, especially tropical seawater (London, 1971; Hubbert, 1973). The water vapor carries heat to the atmosphere wherever it condenses, forming clouds or rain. This is the principal mechanism for heating the tropical and subtropical atmosphere, though surface contact and direct absorption in the air also help, and the difference in weight between the warm subtropical atmosphere and the cold polar atmosphere is the primary driving force of the global air circulation.

Most of the incoming solar radiation is either directly reflected or it heats the ground and is radiated as heat back into space, but the 23 percent that causes evaporation is an ample source for driving the winds that require only a few percent—more than 2 percent and perhaps about 4 percent as already mentioned—even operating as a rather inefficient heat engine. With so much of it directly re-radiated, the incoming sunlight is thus used with only something like 4 percent efficiency to drive the winds. By comparing the total kinetic energy of the atmosphere, 330 billion megawatt-hours, with the power of incoming sunlight, 170 billion megawatts, we see that it would require only two hours of sunlight used at 100 percent efficiency to feed in the amount of energy stored in the motion of the atmosphere. Used at roughly 4 percent efficiency, it would instead take something like two days. Comparing that figure with the 8 million megawatts used by mankind, one sees that, if the kinetic energy would be simply converted to man's use without further input, letting the circulation run down, it would take about four thousand hours or five years to use it up, as already mentioned. The contrast between the two days it would take the sun to put the energy in, through the inefficient process it employs, and the five years it would take man's use of energy to drain it is another expression of how much more than ample is the drive and how little difference to the general circulation man's harnessing of the winds could make.

This makes it apparent that there is plenty of energy to spare in the wind and that the real problem is finding ways to extract it for man's use. The energy is diffused through enormous space and this fact makes it difficult to extract even the very small part of it that could make a substantial contribution to our needs.

Appendix 2

Material Requirements for Large and Small Wind Turbines

The relative cost-effectiveness of wind generation of electric power by various sized wind turbines will depend on several factors contributing to fabrication, siting, and maintenance costs. Both monetary cost-effectiveness, which is apt to be decisive in policy decisions, and energy cost-effectiveness (Odum, 1976), which has ecological as well as fiscal implications, are important. The relation between construction material and power output is important to both, though more important to energy effectiveness than to cost-effectiveness. This factor alone, the question of pounds per rated kilowatt, as a function of scale, is considered here.

In order to isolate this factor, we assume that wind turbines of the same basic design but of widely different sizes are to be built and deployed in the same flat environment, for example in the strong-wind regions of the western Great Plains. This will serve as a basis for comparison, from which deviations arising from different designs for particular sizes and terrain may be considered later.

The basic design of a conventional horizontal-axis wind turbine shall specify the number of blades, presumably either two or three, the ratio of blade span to tower height, and the general format of each. The thickness of the spars, both on the tower and the rotating members, shall be varied with scale to meet stress requirements, the determining stress requirements being those encountered in the maximum plausible storm conditions.

Let the scale parameter for length of spars be Z. This may, for example, be the shaft height above ground. In normal conditions, wind speed V increases with height above ground, but rather slowly, giving a limited incentive for making windmill towers high. The usual assumption (Reed, 1975) for useful wind intensities is

$$V \sim Z^{1/7} \quad \text{or} \quad V = v_o Z^{1/7}. \tag{1}$$

If we assume constant efficiency, the power developed is proportional to the mean cube of the wind velocity (the rate at which kinetic energy, proportional to V^2, is delivered to unit area) and to the area of the disc swept out by the blades, that is, to the square of the linear dimension:

$$P \sim V^3Z^2 \sim v_o{}^3Z^{17/7}. \tag{2}$$

Actually the vertical variation of wind speed is different for a gentle wind and a strong wind, the exponent 1/7 in equation (1) being an average over the useful range. The appropriate exponent varies from about 1/6 in a wind measuring 15 miles per hour at a height of 50 feet to about 1/8 in a 33 mile-per-hour wind and then the exponent drops rapidly to zero at 50 miles per hour and above, meaning that for very strong winds V is independent of Z or about the same at all heights. These numbers are inferred from wind speeds recorded at heights 400 feet and 50 feet over a long period of time at Hanford, Washington. Lacking more direct data for the situation in violent storms, we assume that the maximum plausible wind V_m for which the strength of the structure must be designed, is independent of height.

While Z is a measure of the length of the spars in the structure, we let X similarly be a measure of their thickness. The force of this wind on a unit length of a spar is taken to be proportional to $XV_m{}^2$ (that is, to the rate of change of momentum if the impinging wind is stopped to a certain degree). The force on a given portion of the structure is then proportional to $ZXV_m{}^2$ and the moment of this force about the base, for example, is proportional to $Z^2XV_m{}^2$.

Most of the spars of a tower structure may be under either tension or compression, depending on the direction of the wind. The greater demand on their strength generally occurs when they are under compression, so this will determine the spar thickness, proportional to X. The maximum compressive load L_c that a spar can bear before becoming unstable and likely to bend or crumple is proportional to X^4/Z^2:

$$L_c \sim X^4/Z^2. \tag{3}$$

(This is obtained by considering a slightly bent beam of constant curvature, C. In a lateral cross section through the center of the beam, the elastic or Hooke's Law force per unit area at any point is proportional to CX, thus the force in each element of area to CX^3 and the moment of this force about a neutral point to CX^4. This is equated to the moment of L_c about the center due to a lateral displacement of the ends relative to the center, proportional to CZ^2, to give $L_cCZ^2 \sim CX^4$.) Since the overall design format is preserved, Z is a measure of the width of the base as well as of the height of the tower. The bending strength of the tower about its base is then proportional to L_cZ and with this equated to the moment about the base exerted by the maximum wind we have

$$L_cZ \sim X^4/Z \sim XZ^2V_m{}^2 \qquad \text{or} \qquad X^3 \sim V_m{}^2Z^3. \tag{4}$$

Thus we arrive at the simple, though not at first glance obvious, result that the thickness scales directly with the height, $X \sim Z$, and one could use the same design drawings with a change of scale. For the blades and their supporting spar the simple calculation is slightly different but the result is the same.

The tower designed with this criterion is presumably several times as strong as needed to hold aloft the weight of the turbogenerator structure. Thus this simple scaling should be close to the actual one.

The mass of material required in the construction is

$$M \sim X^2Z \sim V_m{}^{2/3}Z^3. \tag{5}$$

Combining this with equation (2) we find that the material requirement per unit power output is

$$\frac{M}{P} \sim \frac{V_m{}^{1/3}Z^3}{v_0{}^3Z^{17/7}} \sim \frac{V_m{}^{4/3}Z^{4/7}}{v_0{}^3} \sim \frac{V_m{}^{4/3}P^{4/17}}{v_0{}^{63/17}}. \tag{6}$$

As an approximation sufficient for these rough scaling purposes one might simplify the fractional exponents thus:

$$\frac{M}{P} \sim (V_m{}^{2/3}/v_0{}^4)P^{1/4}. \tag{7}$$

Thus, if judged solely from the material requirement standpoint, a smaller wind turbine is more economical than a large one. Within one narrow range of power ratings, the difference arising from the factor $P^{4/17}$ is not great: a 2-megawatt machine has a mass per unit power 18 percent greater than a 1-megawatt machine. Between large commercial power machines and small home machines the difference is appreciable: between 1 megawatt and

1 kilowatt the factor is about 5. Another example: it would take 72 percent more material to build one 2-megawatt turbine than ten 200-kilowatt ones and from this materials standpoint alone one would select the latter to supply high-line power. However, engineering studies by Putnam and his associates (Putnam, 1946) have concluded on the contrary that the most economical sizes for this purpose are in the range about 1.25 to 2.5 megawatts, with a flat cost minimum varying only about 2 percent over this range. The materials requirement presumably dominates in making the rise toward higher powers and various economies of size can cause the rise toward smaller powers, such as unit construction costs, siting costs, maintenance costs, etc. In the mountain location considered by Putnam, siting costs may dominate his estimate of the reduced economy of smaller machines and it is possible that material savings might dominate and make fractional-megawatt wind turbines more economic in flat country.

Scale is even more inadequate as a sole criterion of the material use of large commercial machines as compared with small home-size machines. Equation (2) may be rewritten

$$Z \sim P^{7/17}/v_o{}^{21/17}. \tag{2'}$$

If a 1-megawatt machine is 150 feet high, for example, this could make a 1-kilowatt machine in the same wind regime $150/10^{21/17} = 9$ feet high. Even in isolation on the Great Plains, where this would be applicable, one would want to go higher to put the lowest sweep of the blades above the height of a man. But in most residential situations one would go higher still to reach above trees and buildings, perhaps to a height of 30 feet. This scaling up of the height of the tower by a factor 30/9, for example, would increase its mass by a factor $(30/9)^3 = 37$. The tower before this increase is only about a quarter of the total mass (if we judge by the Smith-Putnam pre-production design, whose tower weighed 90 out of a total of 352 tons) and increasing its mass by 37 increases the total mass by a factor of 10. Thus, rather than to have a mass per unit power five times less than the large machine, the 1-kilowatt machine on a 30-foot tower has a mass per unit energy about twice as great as a 1-megawatt machine 150 feet high.

The scaling discussed here can be expected to apply approximately to the tower, blades, and perhaps also to the hub assembly, which together account for about 60 percent of the weight of the Smith-Putnam pre-production model. The other 40 percent com-

prises largely generator, gears, and their housing in the "pintle
assembly," which are items that would be expected to have a mass-
per-unit-power ratio more nearly independent of P. Thus the variation of the total with P is probably somewhat less pronounced
than here indicated, but these considerations serve to give a rough
preliminary indication of the variation.

In practice the dependence on the wind regime, represented by
the parameter v_o in equation (7), is also an important criterion in
judging the relative materials demands of small and large wind
turbines. The large machine producing commercial electric power
can take advantage of a favorable choice of wind regime whereas
the small machine for a particular household cannot. A survey of
wind regimes in the United States (Reed, 1975) indicates a
variation from around 100 watts/meter^2 in many areas to over 300 in
favorable areas such as southern Wyoming and the Texas and
Oklahoma panhandles. Between these extremes, v_o^3 differs by a
factor 3. Thus a factor of about 2, as an average, in M/P could
be deduced from this consideration in favor of the large machines.

In the economic decision-making process there is no direct competition between small home wind turbines and large commercial
ones. The incentives for making the investment and undertaking the
upkeep are quite different on the household and on the commercial
scale and materials requirement will be only a secondary consideration in either case. But in the course of national energy planning,
one may envision incentives encouraging either the household or the
commercial scale. The question of net energy is important, in other
words, the question of how long it takes a wind turbine to deliver
as much energy as is required to manufacture it. Here mass per
installed kilowatt is an important criterion and appears to give the
large-scale installation an advantage (unless transmission lines,
which have not been considered here, becomes excessively long).

Combining the factors discussed here, 1/5 from straight scaling,
10 from extending the small machine above buildings, and 2 from a
more favorable wind regime, we arrive at a factor of about 4 in
favor of the large-scale commercial venture, but this still has to be
modified (in some cases by as much as 25 percent), by an unfavorable factor to account for the transmission lines in the large commercial application.

It is quite a different matter to decide whether commercial
electric power on a large scale should be generated by many
rather small wind turbines or fewer large ones. Here we might
compare, for example, the deployment on flat land of ten thousand
1-megawatt machines with the deployment of a million 10-kilowatt

machines generating the same total power. We think of them in a square array, each in a corner of the squares of a checkerboard. The distance between them, the "lattice distance," is determined by the consideration that each turbine casts a wind shadow that is dissipated in the distance it takes for air from a higher level to mix with the air near height Z as well as horizontally behind each machine. As has sometimes been suggested (Reed, 1975), we take the lattice distance to be 10 Z. (There is a Canadian study [Templin, 1974] that suggests a spacing of 30 Z but it is based both on a two-dimensional model and on the questionable assumption that the mixing induced by an array of wind dynamos is the same as would be induced by a uniform distribution of low bushes.) Incidentally, the spacing 10 Z would be about a third of a mile for 1-megawatt machines 170 feet high, say, making 9 turbines per square mile so that 10,000 turbines would require about a thousand square miles, one-sixth of the area of the slender Oklahoma panhandle.

The factor $P^{7/17}$ in equation (2′) means that the smaller 10-kilowatt machines, less powerful by a factor of 100, are shorter by a factor $100^{7/17} = 6.7$ and the lattice distance is correspondingly shorter. The area of each square, or the land area per turbine is smaller by a factor $6.7^2 = 45$ for the smaller turbines and the area of the entire array is thus $100/45 = 2.2$ times as great for the small machines generating the same total power. The total materials requirement for a hundred times as many smaller machines is down by a factor $6.7^3/100 = 3$. Here we assume that the smaller machines are simply scaled down from the larger ones so the lower sweep of the blades may come down to about the height of a man on horseback. It would then be desirable to have a secure fence around each tower but this would interdict only about 1 percent of the land from agricultural or grazing use.

From the point of view of net energy and use of mineral resources, these simplified considerations make the use of relatively small wind-electric turbines attractive, requiring only about a third as much material at the expense of about twice as much land area if the power would be generated by a hundred times as many smaller machines. The greater fabrication and deployment costs of the much larger number of smaller machines may, however, offset this advantage when it comes to dollar-per-installed-kilowatt costs, in keeping with the indications already mentioned that the most economical size is in the 1.25- to 2.5-megawatt range. Maintenance costs which go beyond the installed-kilowatt costs presumably also favor the fewer larger machines.

The ultimate economic criterion will not be installed-kilowatt cost but rather cost per average generated kilowatt and this is sensitively dependent on cut-in speed of the turbines. Smaller turbines tend to have the advantage of lower cut-in speeds but this is partly compensated for by the fact that in a given wind regime the smaller turbines are subject to the slower wind closer to the ground. The balance here is dependent on the average wind, the smaller machines with their lower cut-in speed being favored in less windy regions.

The Power a Wind Turbine Can Extract from the Wind

One might picture the maximum energy available to a wind turbine as the kinetic energy of the wind passing through an area A equal to that of the blade circle. This cannot all be extracted because the wind has kinetic energy as it leaves the downwind side of the blade circle. As an idealized and somewhat inconsistent model, consider a cylinder of air of area A entering from far upwind with the wind speed V_1, passing through the blade circle with velocity V and leaving far downwind with velocity V_2. For this model Betz (1920) has quite simply derived an upper limit of the power extracted by the wind, showing it to be 16/27 or 59 percent of that incoming kinetic energy.

The mass flux of air of density ρ through the blade circle is $M = \rho A V$. From conservation of momentum, the force on the blade circle is $F = M(V_1 - V_2)$. From conservation of energy the power expended on the blade circle is

$$P = (M/2)(V_1{}^2 - V_2{}^2) = FV = MV(V_1 - V_2) = \rho A V^2 (V_1 - V_2).$$

From the second and fourth members one has

$$V = (V_1 + V_2)/2 \qquad \partial V/\partial V_2 = 1/2$$

and from the last member, keeping wind speed V_1 constant, one may find the V_2 for which the power is a maximum:

$$\partial P/\partial V_2 = 0 = \rho A[V(V_1 - V_2) - V^2]$$
$$V_1 - V_2 - V = V_1/2 - 3V_2/2 = 0. \quad V_2 = V_1/3. \quad V = (2/3)V_1.$$

Then the maximum power is

$$P = \rho A V^2(V_1 - V_2) = \rho A V^3 = 2(2/3)^3 \rho A V_1{}^3/2$$
$$= (16/27)(\rho A V_1{}^3/2).$$

The changes in velocity occur gradually with a gradual increase of pressure above atmospheric on the upwind side of the blade circle, a sudden drop of pressure at the blade circle, and a gradual increase of pressure to atmospheric on the downwind side. The power is derived from the pressure drop at the blade circle. A more nearly consistent model considers an expanding tube of air having a cross section A_1 and velocity V_1 far upwind and A_2 and V_2 downwind, with $A_1V_1 = AV = A_2V_2$. The boundary of the tube curves outward on the upwind part, pushed outward by greater internal pressure there, and similarly inward downwind. Some further thought shows that the derivation applies also to this model, the total pressure force on the segment between A_1 and A_2 being zero.

Reality of course differs from the model in several ways. One way that increases the theoretical maximum power is the mixing across the boundary of the tube, energy and momentum from beyond the area A helping to push along the slower air in the tube. A difference reducing the limit is that a conventional single rotor imparts a spiral motion to the downstream air, but the kinetic energy that is wasted in rotation of the stream amounts to only a few percent of the power output if the tip speed ratio is high, 6 or so. This percentage is in rough approximation equal to the power ratio, say 16/27, divided by the square of the tip speed ratio.

There is thus no reason in principle that the Betz limit of 59 percent should not be exceeded in practice with good blade design and high tip-speed ratio, though measured values are mostly in the neighborhood of 50 percent or less.

Appendix 3
Reactor Safety Studies

Some students of reactor safety conclude that it is not only extremely difficult but in principle impossible to determine the likelihood of calamitous accidents, short of having them actually happen. Dr. Clifford Beck, AEC director of regulation, in an internal memorandum to the commission on the occasion of a 1965 reactor safety study, wrote, "Here is encountered the most baffling and insoluble enigma existing in our technology: it is in principle easy and straightforward to calculate potential damages that might be realized under postulated accident conditions; there is not even in principle an objective and quantitative method of calculating probability or improbability of accidents or the likelihood that potential hazards will or will not be realized" (Kendall 1977, pp. 1–10).

Since the decision was made in about 1954 to go ahead with a large government-sponsored industrial power reactor program, it has been necessary either to make reactors safe or at least to provide credible assurances that they are, that is, to attain either successful development or successful promotion, or both. The AEC and it successors have attempted both and it is not clear that they have succeded in either.

Quite apart from the continuing developmental efforts on safety measures, there was a series of three official AEC studies related to reactor safety. The first step was straightforward enough: recognizing that an accident releasing much of the radioactivity in a reactor core could happen, even if it is very unlikely, the AEC put the question to a study group centered at Brookhaven National Laboratory: if a large release should occur, what would be the consequences? The result of the study was contained in a 1957 report known as WASH-740 or the Brookhaven Report. It concludes that release of half of the radioactivity in the core of a reactor of that day (about one-sixth as large as recent reactors), located eighteen miles from a major city, might cause about three thousand deaths and do about seven billion dollars worth of property damage.

The legislative reaction to this conclusion, as the principal officially established fact about reactor hazards, was a peculiar compromise, the Price-Anderson Act. The logic of it seems to have been this: one need do nothing about the deaths but since an accident could do great property damage that can be given a dollar value, there should be insurance to cover it. Since, however, the accident is presumably very unlikely, the insurance need not cover all of the estimated potential damage, but perhaps about one-tenth of it. Since the magnitude of the potential damage is so much greater than from other industrial hazards and there is no way to judge the likelihood of an accident leading to claims much larger than the insurance coverage, the nuclear industry should be absolved of responsibility for such great damage, that is, legally declared in this respect irresponsible. Incidentally, industrial insurance companies were unwilling to assume much of the risk so the government has to provide most of the limited coverage up to the one-tenth level, and the public is left without recourse for the rest. With this protection, the various members of the reactor industry felt confident enough to proceed to install reactors that they declared to be safe. This was the first step in providing some assurance about reactor safety.

The next attempted step was the commissioning of a study on reactor safety leading to a 1965 internal report known as the WASH-740 update. This again did not attempt to judge the likelihood of occurrence of serious accidents, as indicated in the above quotation from Dr. Beck, but confined its attention to reassessing the consequences of accidents in larger and more modern reactors than considered in the original WASH-740 report. The result of this presumably more careful study was quite unexpected, namely, that the consequences could be considerably worse than anticipated in the original report.

The question then arose what to do about such an adverse result and the answer to this question is perhaps the most damaging single bit of evidence concerning the credibility of the official assurances about reactor safety. The conclusions of a steering committee meeting on the revision of WASH-740 were reported in an internal 1964 memorandum that was not made public until much later after litigation that followed passage of the freedom of information act (Kendall, 1977, app. A, p. 148). Its summary stated, "The results of the study must be revealed to the Commission and the JCAE without subterfuge although the method of presentation to the public has not been resolved at this time."

"The results of the study suggest that the Price-Anderson liability level should not be reduced. Rather, an increase by a factor of 40 is suggested by the calculations (280 billions)." The subsequent decision was to suppress the report.

The next step was the commissioning by the AEC of a much more extensive study reported as WASH-1400, the Reactor Safety Study. The study involved about sixty people. About a third of them at AEC headquarters near Washington did most of the work but depended on inputs from many outside consultants. The official director was Norman Rasmussen, professor of nuclear engineering at MIT, and the report sometimes bears his name.

The report is technically competent and even helpful in some special ways but is seriously misleading on the overall question of nuclear safety. There is evidence that the study was set up in such a way as to obtain the desired result and to substantiate a declaration that reactors are safe. In a 1973 memorandum within the AEC, E. Gilbert mentioned as one of the disadvantages of carrying out a reactor safety study that "the facts may not support our predetermined conclusions" (Shapley, 1977). In an internal letter, later made public, proposing the nature of the study, Professor Rasmussen (who was later made its director), wrote that major parts of the study would "be a manageable task that might have significant benefit for the nuclear industry" and that "the report to be useful must have reasonable acceptance by the people in industry" and further that great care must be taken in organizing the study because "once we start, our results may become public knowledge" (*New York Times,* April 27, 1977). This seems to imply that those aspects of nuclear reactors that might turn out not to appear safe should be excluded from the study from the beginning, as is indeed consistent with some of the omissions from the report (Kendall, 1977; Webb, 1975). Even with the best of intentions, the inevitable choice of competent talent for the study largely from among professionals in the nuclear enterprise would introduce a bias toward the presumption of sufficient safety.

While reactor safety has been studied long and assiduously in national laboratories, WASH-1400 probably represents a more systematic study of the sequences of things that might go wrong ever to be assembled in one place. Its identification of some of those sequences of single failures making the leading contribution to risk, for which its methodology is well suited, may lead to improvements in reactor design and practice. In this way, at least, the report has some value. It concludes that the only significant chance of reactor calamity is by way of core melting and the only way considered plausible for the molten core products to reach the atmosphere is by way of melting through the bottom. This conclusion depends on confidence that pressure vessels will not burst. The likelihood of a core meltdown is given as one chance per 17,000 reactor years, or a bit less than 10^{-4} per reactor year. That would mean that with a thousand reactors in the future there might be a core meltdown once in about seventeen years. This is somewhat more often than was previously thought. However, the report goes on to adduce reasons why a core meltdown may be considered almost benign, and why it is very unlikely that many people will be killed or injured even if the core does melt through the bottom.

In brief summary, it is considered that the radioactive release will be sufficiently limited by filtration through soil to avoid producing a lethal dose beyond about twenty miles; that only about one reactor in a hundred will be close enough to a large city to do great harm; that the wind will be blowing toward the city about one-tenth of the time; and that only about one-tenth of the populace of the affected region will be seriously exposed, some people having been evacuated and others partially shielded by buildings. It is also noted that, for a large radioactive release, the heat generated by the radioactivity will be sufficient to lift the radioactive plume somewhat off the ground. This summary is inadequate to convey the complexity of the treatment, but gives the general idea.

The report in preliminary form, thousands of pages that take up nine inches of shelf space, was released in August, 1974, and in the final form in October, 1975. But already in January, 1974, high AEC officials started making public statements using the purported conclusion of the report, that reactors are safe, in their promotion of the rapidly expanding reactor program. This was half a year before the public had any way of knowing the hypotheses on which the claim of safety was based. Thus the report was politically misused even before it was issued.

When the preliminary report was issued, after two years of work and an expenditure of about three million dollars, the public was invited to submit critical comments within two months to facilitate making the final report less open to criticism. Several groups of scientists did undertake critical studies and submit serious reports, in no more than preliminary form by the time of that early deadline, but in good shape a few months later. Of these we shall cite in particular the studies sponsored by the American Physical Society (APS, 1975) and by the Union of Concerned Scientists (UCS) and Sierra Club (Kendall, 1977).

These reports were in many respects highly critical of the main conclusions of WASH-1400, finding those conclusions to be optimistic, by factors as high as forty and in some cases much more, about the likelihood and consequences of severe accidents. Yet the nuclear promoters were able to evade this obstacle and achieve one important political purpose of that official report, to obtain renewal of the Price-Anderson (or "Nuclear Irresponsibility") Act, by denying the existence of opinion contrary to the findings of WASH-1400. This artful maneuver involved timing the political action just after the appearance of the final form of the report. Indeed, the principal critics had not even been able to obtain copies of the final report, despite requests, before the following exchange took place in a hearing before the Joint Committee on Atomic Energy:

Congressman McCormack: To summarize, there is no substantive scientific group anywhere with whom you find lingering differences?
Professor Rasmussen: None that I'm aware of.

Congressman Price (the surviving member of Price and Anderson), in arguing that there was no further reason to delay voting on the Price-Anderson Act, testified similarly before the Rules Committee:

The final report was almost identical to the original version. The only difference being one or two minor areas where they got together with the people who raised exceptions to figures and there was a final meeting of the minds.

Thus the final report, too, was politically misused. The final report did include quite a lot of changes, including some increase of the risk estimates and an additional appendix XI on "Responses to Comments on the WASH-1400 Draft," but no major changes of its conclusions. The inadequacy of some of these responses and how far they remain from representing a "meeting of the minds" will become apparent in the following discussion. But quite independent of any judgment of the content of the report and the adequacy of the responses, the timing of Professor Rasmussen's representation to the committee, before the final report had even been seen by the outside scientific critics, raises legitimate doubts about his credibility and, by inference, about the credibility of the report prepared under his direction. The report is too massive to be summarized briefly but the following description of some of its content, by way of discussing some of the criticisms, may help convey an appreciation of the nature of reactor safety problems.

WASH-1400 pursues in considerable detail the probability and likely consequences of some of the less unlikely types of accidents, primarily those the reactors are required to be designed to handle, while other presumably more unlikely types are judged, some of them quite arbitrarily, to be too improbable to contribute significantly to the overall risk. Some criticisms focus on the former, on both the likelihood and the likely consequences, while others challenge the judgment that certain accidents are so unlikely that they should not cause concern and furthermore challenge the presumption that all likely paths to disaster have been imagined.

One subject of contention is the probability of failure of the main pressure vessel, whether merely leaking very rapidly or bursting wide open to cause a meltdown of the core, penetration of the outer containment, and having dire consequences beyond those considered in the report. The very extensive experience with high-pressure vessels in non-nuclear industry includes some failures, and these indicate a probability of primary pressure vessel failure of about 10^{-5} per reactor year. WASH-1400, however, considers that reactor vessels are made so much more carefully than ordinary commercial vessels as to be ten times less likely to fail at all and further, that only one-tenth of the failures will be bad enough to have disastrous consequences. On this assumption it introduces a factor one one-hundredth and gives the probability of a disastrous primary pressure vessel failure as 10^{-7} or one ten-millionth per reactor year. The critics point out that there are other factors in the opposite direction; that it is more difficult to ensure the integrity of welding seams in the thicker reactor vessels and to inspect them in the presence of radioactivity, and that the role of radioactive embrittlement of metal is not reliably known. The response to this (WASH-1400, p. XI 8-1) is that all failure data have been taken into account and that "although there is some opinion in the United Kingdom that the probability of catastrophic failure of the reactor pressure vessel should be about 10^{-5} per reactor year, the study does not believe that this value is very realistic."

The response goes on to state that even if the probability of gross vessel rupture large enough to negate protective measures and cause core melt, were as high as 10^{-5} per reactor year, it would then just begin to contribute to the overall risk (in particular of "large-consequence accidents") and would not change the results of the study (WASH-1400, pp. 63 and XI 8-1). How this contribution is figured is not explained but it may include some optimistic factors minimizing the effects of the release on the population. The "results of the study" are, for large-consequence accidents, that the probability of an accident bad enough to cause 3,300 early fatalities and 45,000 early illnesses is one in a billion reactor years ($10^{-9}/RY$) and, at the other extreme for low-consequence accidents, the probability of one causing perhaps one early fatality and 300 early illnesses is one in a million reactor years ($10^{-6}/RY$) (Wash-1400, p. 83). The figure 10^{-9} includes a factor 10^{-3} for the probability that the weather and population distribution will be favorable for this catastrophe following a delayed breach of containment that might be caused by a steam explosion when the molten core falls into water. This factor would not apply to the more intense release from a spontaneous pressure vessel burst, since there might be no delay between the end of the chain reaction and the release. "Illness" here of course means radiation sickness, with its extreme unpleasantness and lingering debilitating effects.

Incidentally, the twelve-page "Executive Summary" of the WASH-1400 report, which seems to have the purpose of informing busy decision makers very briefly, avoids mentioning even these low estimates of fatalities and early illnesses by presenting as the result of the study a table of the "most likely consequences of a core melt accident" which are essentially zero. This seems to imply there need be no concern for such effects. When read carefully, it says that, of all core melt accidents (which are predicted to happen once in 17,000 reactor years) most are almost completely contained (by the soil into which the core descends) and only a minority of them release enough radioactivity to have serious effects. It is of course that harmful minority in which executives should be interested, but in the summary only the almost benign majority, those having the most likely effect, are tabulated. Another table does give the claimed average expectation from operation of 100 nuclear plants as only two fatalities and twenty injuries per year compared with eight fatalities from lightning and much larger numbers from other hazards such as automobiles.

The American Physical Society report (APS, 1975) was particularly critical of the estimates in the WASH-1400 draft of health effects from the radioactivity released in a serious accident and, indeed, some of the outright mistakes it uncovered there were at least partially corrected in the final WASH-1400 report. The draft considered the effects of ground deposition of radioactivity only for the first day on the assumption that the most heavily contaminated areas (within twenty miles) would be evacuated after that, but actually the long-term cancer deaths would be much more numerous in the larger surrounding areas and the APS study found the effect increased by a factor 25 when more properly estimated. In taking into account this error, a more exact sensitivity calibration, and the effects of the lung dose and beta-ray–induced thyroid cancer neglected in the draft, the APS report indicated that the estimated number of cancer deaths should be increased by a factor of about 40 over the draft WASH-1400 claims, making about 13,000 cancer deaths rather than 310 from the specific serious accident considered, as well as about a factor 25 for genetic damage. Yielding a bit to this criticism, the final WASH-1400 report lists increases over the draft by a factor 1.3 for early fatalities, 14 for latent cancer fatalities, and 1.6 for genetic effects.

While there is a great deal of discussion in these reports on those aspects of reactor safety—both of the likelihood of specific accidents and of their likely consequences—that can be considered in semiquantitative fashion, there remains a serious question whether the risks thus discussed may not be overwhelmed in reality by greater risks from other mechanisms that are inadequately discussed and summarily dismissed or not even mentioned at all. Such mechanisms include power excursions (going well beyond prompt critical), autocatalysis and common-mode failures. More generally stated, the concern is whether, after searching with the elaborate event-tree analysis which is very useful for eliminating physically meaningless sequences of events, one has really been able to discover all possible paths to calamity and then properly assess them. Experience with failures from unexpected causes in other fields, such as the Tacoma Narrows Bridge collapse and the East Coast blackout, as well as in the nuclear field, such as the Fermi reactor partial meltdown and the Brown's Ferry fire, suggest that one cannot be confident of having thought of all possibilities.

A related haunting question is whether it makes practical sense in accord with experience to do what is theoretically reasonable and obtain an extremely small probability of a multiple failure by multiplying together several small and rather uncertain probabilities of the individual failures. Consider, for example, the probability that a pipe rupture in the high-pressure system of a pressurized water reactor will cause a core meltdown and subsequent specified extensive casualties. WASH-1400 lists five successive things that have to happen (WASH-1400, p. XI 3–37). First the pipe must rupture, then a failure of the protection systems that should prevent core melt, followed by a breach of the outer containment when high pressures develop, then the weather must be right for carrying the radioactivity to the population, and then the population distribution must be right for many people to be exposed. The probabilities for these are listed as follows:

Pipe rupture per reactor year	10^{-3}
Systems failure	10^{-2}
Containment failure	10^{-1}
Weather conditions	10^{-1}
Population distribution	10^{-2}
Entire event per reactor year	10^{-9}

or one chance in a billion per reactor year. This is typical of the way the probabilities reported in WASH-1400 are calculated.

The UCS-Sierra Club study emphasized the unreasonableness of the small probabilities of disastrous events in WASH-1400 by citing several cases of serious but not disastrous accidents that have actually happened even though they would be estimated a priori by the general approach used in WASH-1400, to be virtually impossible. In one of these the Dresden II boiling water reactor near Chicago went spectacularly out of control for two hours in a mad sequence of component malfunctions and operator mistakes. The initiating event was merely the malfunction of a gauge that showed the pressure to be much too low, so the operator compensated by increasing the pressure. In the subsequent confusion of rising and falling pressure and temperature in attempts to compensate, with readings or lack of readings from faulty instruments, two experienced operators concurred in deciding to ignore regulations in the emergency. Only one pressure gauge was still working and the pressure went far beyond its range, so it is not known how close the excessive pressure came to bursting the containment. By multiplying together probabilities of individual malfunctions as is suggested in WASH-1400, the USC study finds the probability of this accident to be something like 10^{-37}, billions of billions of times less likely than the least likely accident estimated by WASH-1400. This leaves plenty of leeway to be more generous with the methodology and the individual probabilities and still come out with a virtually impossible accident that happened.

Another incident cited by the USC occurred in the Oak Ridge Research Reactor and was reported (Eppler, 1970) by an engineer there specializing in reactor safety as being "almost unbelievable." In order to make very sure that a certain important protective system would not fail because of power failure, three separate and independent storage-battery powered systems were provided in three separate rooms, a case of threefold redundancy that is common in reactor safety systems. Yet the reactor operated for some hours without this protective system, when it was fortunately not called upon to perform, because all three storage batteries were dead. This happened because of twenty-one individual failures, the absence of any one of which would have avoided the incident. These twenty-one are grouped as seven sets of three failures each that happened for the same reason although they were supposed to be independent, seven triple "common-mode failures," as they are called. Calculated by methods suggested in WASH-1400, these seven sequential common-mode failures, according to the USC study, had individual probabilities of 10^{-5}, 10^{-2}, 10^{-5}, and four more 10^{-2} each, or even less, giving the overall a priori probability of the incident as 10^{-20} or even less (per reactor start-up, which is not very different from per reactor year). Even if the question of the common-mode failure methodology is eliminated by very generously taking a single-failure probability in place of each triple-failure probability, the overall probability still comes out to be 10^{-15}, or one chance in a million billion for this incident that happened, whereas WASH-1400 considers that severe accidents are extremely unlikely that it assesses to be a million times more likely than that (10^{-9} per reactor year).

In considering the possibility of a sequence of events each with an estimated small probability and each probability but the first conditional in the sense that it comes into consideration only *if* the previous event has occurred, the theoretical presumption is that the total probability is the product of the many small individual probabilities. The validity of WASH-1400 depends entirely on this seemingly reasonable presumption, yet it just does not hold up in practice. In this light the small theoretical numbers of nuclear fatalities that WASH-1400 compares extensively with observed numbers of fatalities from other hazards seem to have no validity at all.

The WASH-1400 comment and response to this and similar criticisms are as follows:

COMMENT 3.2.10
Examples were given of actual incidents that involved several sequential human or equipment failures. The comment questioned the ability of the study to predict such events using the methodology employed in WASH-1400.
(Union of Concerned Scientists, The National Intervenors)

RESPONSE
In performing its assessment, the study reviewed not only the examples cited in the comment but also many other sources of pertinent data. The study's analyses were not meant to be taken out of context and extrapolated to different situations or different sequences. Sequential failures must be treated by sequential methods; alternatively, it is necessary to identify, by the use of methodology similar to that discussed in sections 3.1.2.1 and 3.1.2.2c of this appendix and in appendix 1, single based causes that govern the sequences of failures. In one instance cited, aging was used as an example of a common mode failure. It should be recognized that the study did not include extreme aging considerations since the applicability of its results is limited to only the next five years.

The first reply amounts to saying "we didn't intend you to test our analyses in other situations where it is possible to compare with known results.'' The situations just cited were indeed examples of sequential failures treated by sequential methods. In the Oak Ridge incident of seven triple failures, for example, there was a probability of an initiating event, which was the throwing of three wrong switches, and a conditional probability for the next, failure to report the behavior of machinery following from the first event while throwing the wrong switches back, the two probabilities to be multiplied, etc.

But the last sentence of the response is a remarkable reply to a different question, a reply devastating to the basic significance of WASH-1400. As regards the possibility that accidents might be caused by components wearing out, ''the study did not include extreme aging considerations since the applicability of its results is limited to only the next five years.'' How, with this astounding disclaimer sequestered in an appendix, can it be claimed that the study certifies the safety of 100 reactors in the 1980s? One wonders what all the ponderous concern for the validity of WASH-1400 is all about. It actually denies its own relevance.

But quite aside from this admission of a five-year limit on applicability, the failure to consider the effects of aging of the components of a system so subject to deterioration as a nuclear reactor, and even this failure alone, casts grave doubts on the validity of any of the risk assessments of the study.

Appendix 4

Conjectures on the Neglect and Future of Wind Power

From the foregoing it is clear that large-scale wind power presents an important option for meeting a substantial part of our energy needs economically and beneficially soon and yet in government energy planning it has been declared, along with fusion and other prospective new energy sources, to be in need of extensive research and development and to be promising for the twenty-first century but not for this century. High-level policy makers must know that it is technically absurd to put wind power and fusion power in the same readiness category and if they do know it, their doing so must be for the sake of public relations, hoping that the absurdity will not be obvious to the members of the public who know little about either technology.

One may then conjecture to what extent the minor role and unnecessarily slow development program assigned to wind power arises from simple lack of foresight and initiative or from lack of appreciation of the option, or perhaps from deliberate neglect and intentional suppression of wind power in order to keep it from deflecting attention and support from the nuclear enterprise with which the energy policy makers are principally engaged and from which they draw much of their technical advice. If deliberate neglect is an important ingredient in the policy motivation, the present seemingly sensible but unnecessarily cautious development program serves as an effective screen to hide this intention, giving the public the impression through periodic news releases that the reasonable progress it expects is being made in the development of large-scale wind power.

It is administratively convenient, as well as useful for conveying this impression to the public, to lump all the prospective new energy sources together (solar-related, geothermal, and fusion), and to say of the group as a whole that protracted R&D is needed be-

261

fore extensive use and then to divide up the R&D funds equitably among them with the exception that fusion gets most as the senior program.

Thus the perhaps disruptive notion that wind power is past the R&D stage and ready for detailed design and deployment may have been conveniently swept under the rug. The working-level administrators recruited to specialize in wind power would then have found that they were expected to organize a program consisting largely of R&D and thus to spread funds out over many reasonably relevant unanswered questions. Even if they became enthusiastic about wind power they would be motivated to avoid risking any conspicuous mistakes in large-scale demonstrations for fear of losing favor for themselves and their limited program.

It is somewhat gratifying that the program has shown more vigor than might be expected under such limiting circumstances. In spite of the six years the program is taking to reach a megawatt-size demonstration such as was achieved in less than two years in the Smith-Putnam effort long ago, the 100-kw machine of 1976 is being followed within three years by three almost identical wind turbines upgraded in power for windier sites (fig. 55), as well as the first megawatt scale machine of the federal program at Howard's Knob. The favored expectation for the next one of the series is for a 2.5-megawatt wind dynamo designed for a lighter wind regime and having a blade span of about 300 feet (fig. 56). It is discouraging that they are all of the same type, each with two blades downwind from the tower like the Smith-Putnam machine, and that no greater variety of design has been demonstrated. This is the most economical way to compare different sizes but may also represent excessive fear of making a mistake. Instead it may actually perpetuate a mistake, perhaps because two-bladed machines are more prone than three-bladed machines to encounter vibration troubles, for example.

One glaring shortcoming of the program is its persistent rejection of the opportunity to demonstrate offshore wind power in order to exploit the best available winds. This seems to have been motivated partly by excessive fear of making a conspicuous mistake early in the program and partly by the administrative requirement which sets a slow pace by permitting only one large-scale demonstration at a time. After the AEC-ERDA wind-power branch for years resisted private urging and proposals to investigate floating wind dynamos, it was faced with a congressional directive to do so. The directive itself was watered down by the intervention of a strong nuclear proponent in Congress, as has been mentioned, but still required a prompt feasibility study before submission of a demonstration pro-

Fig. 55. The DOE-NASA 200-kilowatt wind dynamo that generates power for community use at Clayton, New Mexico. It is upgraded for stronger winds from its 100-kilowatt predecessor at Plum Brook, Ohio, from which it differs by having stronger blade roots, a stiffer yaw drive to avoid resonance between yaw vibration and rotor speed, and a tower with tubular rather than angular spars. Their design is patterned partly after the German machine of figure 10. The blade span is 125 feet, hub height 100 feet and rated wind speed 18 miles per hour. In 1978 this machine is the biggest success of the DOE wind power program. Duplicates are to be built on Bloch Island and in Puerto Rico.

posal to Congress. This provided a classic example of the possibility of killing a project, if so desired, by choice of the circumstances of an investigation. In a project requiring such innovation of design, it would have been appropriate, if a favorable or fair result were desired, to have the study carried out by at least two teams, including one with a leader known to be both able and enthusiastic about the project.

Actually, a single feasibility study contract was let to one of the nation's largest builders of nuclear power plants. Westinghouse.

Fig. 56. Artist's concept of the 2.5-megawatt wind dynamo being designed by Boeing for DOE as MOD 2. It is intended for use by electric utilities in regions with only moderate wind speeds averaging 12 or 14 miles per hour, though its rated wind speed is 18 miles per hour. Its blade span is 300 feet, much larger than any yet attempted. An early artist's concept of the Plum Brook machine also showed such a slender tower which was, however, replaced by a trussed tower as in figure 55. If this be the shape of things to come in the federal program, it will be of great interest to compare it with about as large a version of the rotating-tripod design shown in figure 57. (*Courtesy The Boeing Company.*)

While the result is not yet known as this is written, if it is unfavorable it will naturally be subject to the suspicion that it might be another case of obtaining a desired result by not having exercised care to avoid a biased investigator. Though it is probably unrelated to this case, there is such suspicion of some practices in other branches of AEC-ERDA, notably the choice of personnel for the Reactor Safety Study leading to WASH-1400, as noted in appendix 3.

In the pandemonium of national policy making it may be that no single motivation is ever dominant, that purposeful suppression of wind power has played a role but only as one of a number of tangled threads in the process and that policy may change drastically and unpredictably. It may be that the geographic and mechanical enormity of a large wind-power array has excluded it from executive imagination when nuclear power seemed administratively simpler, after the technical path had been charted and its industrial organization was at hand and pressing for attention.

The strong but unproved suspicion that there may have been conscious rejection of the large-scale wind power option and neglect of some of its most promising aspects in the past leads to the question of where wind power stands in the present energy planning of the present administration and how it might be introduced into future planning. President Carter apparently came into office with high hopes for some much needed and rather drastic changes in energy planning but with limited experience in the ways of Washington and a need to assemble a top-level team experienced there. This meant going against his reservations about nuclear power and appointing as the top executive the most able available person with high-level energy experience, a strong nuclear advocate and former head of the AEC, James Schlesinger. This implies carrying over into the new administration the old AEC pronouncement that the large-scale wind-power option should be saved until the next century. It leaves the president insulated from the realization that his neglect of the big wind-power option in his energy plan makes a mockery of his claim to be using nuclear power only as a last resort. Substituting a large wind-power program for a substantial part of the projected nuclear expansion would require special initiative by the president at a time when he is running into political opposition from vested interests to some of his more general initiatives in the energy program and is achieving much less of the needed change than he apparently intended.

Since even the transition to a new administration with a newly organized Department of Energy and a new perspective on the challenging problem of future energy has not sufficed to upset established policy and make us opt for large-scale wind power, one wonders if anything will. Perhaps the initiative must come from Congress, but bringing together the disparate interests of Congress on this would probably require some source of political pressure or at least a realization by several key members of Congress from different states that a big wind-power program would create jobs in their districts.

Such pressure and realization is most likely to arise from concerted effort by individual companies interested in the wind-power business. In a sense, this might be a natural outgrowth of present policy providing for the government to demonstrate a few wind dynamos to industry intending that industry will then perceive them to be economic and carry on the expansion of the program from there. Industry might then recognize that demonstration of a larger variety of models would be desirable before gradually going to mass production. They might by then have political clout to induce the government both to broaden its demonstration program and to arrange tax and other incentives for wind power commensurate with those given to other power technologies with which wind power must compete. This would probably be the slow route, taking almost until the next century to be really effective, though it might happen faster. However, the pressure on government to take steps hastening mass production might never develop in this way, for it requires industrial leaders to perceive the economy expected with mass production before the mass production occurs.

Perhaps the greatest hope for rapid progress lies in the vigor of some small independent wind-power companies that might expand their activities from their experience on a smaller scale to marketing units in the megawatt range. Seeing their success, larger industrial and financial centers might perceive the economic viability of large-scale wind power without benefit of federally sponsored mass production.

Three new developments in the period 1977–78 nurture this hope. The most exciting of them in terms of future prospects is the rotating-tripod design of wind dynamo well engineered and successfully demonstrated with a 140-kilowatt model in 1977 at Moses Lake, Washington (fig. 57). It has a broad-based support structure as in the British design of figure 12, but with all three legs of the tripod rotating on a smaller circular track. It is designed to scale to larger sizes and the next one, rated at 3 megawatts in a 40-mile-per-hour wind, is in 1978 being erected for the Southern California Edison Company in a broad pass between two mountains near Palm Springs. There the wind is even better than on the western Great Plains. As was mentioned in chapter 4, its cost and anticipated performance should show even more convincingly than the rest of the discussion there that wind power in windy regions is cheaper than other sources. This also shows the wisdom of looking beyond the single design concept of the federal large-scale wind-power program.

Fig. 57. The Schachle rotating-tripod wind dynamo at Moses Lake, Washington. It is rated at 140 kilowatts in a 26-mile-per-hour wind. It uses hydraulic transmission of power permitting a variable-speed rotor to convert energy efficiently with varying wind speed. This was designed as a prototype for much larger machines. Construction of one rated at 3 megawatts, begun in 1978 under contract with Southern California Edison, was delayed by inadequacy of initial financing leading to design changes under new management. *(Courtesy of Charles Schachle and Sons.)*

Another recent development is the 200-kilowatt wind dynamo (fig. 58) erected by a small Angola, New York, firm on windy Cuttyhunk Island, off the Massachusetts coast (Sheperdson, 1977).

While less directly relevant to American industrial progress, it is remarkable and encouraging that in 1977 the largest wind turbine ever built (2 megawatts) was completed in the little town of Tvind near the North Sea coast of Denmark (fig. 59). Like the pioneering wind generation of electricity in Denmark starting in 1893, this came from the initiative of school teachers, mostly with volunteer labor from three nearby schools and with financing out of teachers' salaries.

The work of small independent companies is thus starting to provide some of the breadth of experience that has been lacking in the federal program. Their business could expand into a much more rapid

Fig. 58. The 200-kilowatt wind turbine, built by WTG Systems on Cuttyhunk Island, Massachusetts, shown prior to installation of the machine cabin housing. It incorporates some of the features of the Gedser Mill of figure 8, but with heavier blades without stays. There is more emphasis on economy than on hydrodynamic efficiency. The blades are of fixed pitch but with rotatable tips for braking. It is presumably economic solely as a fuel saver, supplementing the island's diesel-electric power supply. The diesel's fuel consumption is automatically adjusted according to the strength of the wind.

Fig. 59. The world's largest wind turbine at Tvind, Denmark. It is rated at 2 megawatts in a 33-mile-per-hour wind. It has a 3-bladed downwind rotor 177 feet in diameter on a 174-foot tower, just slightly larger than the old Smith-Putnam machine of figure 6 and with a more refined blade design. Construction was completed in late 1977.

growth in the use of large-scale wind power than is contemplated in government policy, but still not rapid enough to provide a major part of the national energy supply soon. This in itself would be important. It could be much more important as a catalyst for government action to stimulate the growth of a very large, wind-power industry we need.

The reorganization of the federal government effort in establishing the Department of Energy has introduced new administrators at the assistant secretary level, and below, with a variety of attitudes toward policy change. It may be that through their perception, the administration in its effort to balance the energy budget will be influenced both by mounting difficulties with other sources and by a growing appreciation of the viability of large-scale wind power and will soon turn wholeheartedly to the wind-power option.

Wind-power deployment could follow the hydroelectric precedent. There have been two quite different arrangements for government involvement in the nation's problem of generating adequate energy. One follows the hydroelectric model and the other the nuclear model. In the hydroelectric model the government has itself, through an appropriate agency, contracted for the building of the great dams of the West and of the Tennessee Valley and is in the business of generating electric power and selling it to utility companies and industry for distribution. In the nuclear model the government has, again through appropriate agencies and laboratories, itself managed the invention, research, and development of nuclear power and has demonstrated its viability to the point where industry could be urged to take over. This is what is meant by RD&D: research, development, and demonstration. The hydroelectric model has been most actively pursued in the first half of the century, the nuclear model in the second. Perhaps largely because of the recent emphasis on the nuclear model, wind power has been classified with the nuclear model although by its nature it belongs with the hydroelectric model (Hammond, 1977d). It does not need to be invented although the federal wind-power program almost seems to be directed at re-inventing it and ends with demonstration. Wind power's needs for RD&D are minimal and the emphasis should now be on another "D": deployment. Deployment is what the hydroelectric model has achieved for hydro power and should now be doing for large-scale wind power.

When nuclear power was first available to the extent of 5 megawatts at the Argonne National Laboratory in the 1950s, this was more than enough for laboratory needs but the local utility company would not agree to using the surplus on the commercial grid. There was resistance to the government's intrusion into the private sector, usurping what was considered to be an industrial prerogative in the power field even though industry had not been capable of making the original nuclear development. Such resistance was probably responsible for establishing the pattern of the nuclear model, ending with demonstration. Hydroelectric deployment was able to overcome such resistance politically probably partly because it is so different from the central-station power generation to which the utilities are accustomed. Federal deployment of large-scale wind power should be acceptable by the same token. As in the case of the great dams, the government would generate the power from the wind but private investment in the utilities would profit by distributing it.

When deployed in this way, large-scale wind power would be different from hydroelectric power in that wind power is, from our present perspective, practically unlimited in scope whereas the favorable sites for large hydroelectric dams are relatively few and have long been developed by private as well as government initiative. In the case of wind power, the government deployment effort could not only make a substantial contribution to our power needs but at the same time break ground for later large-scale commercial initiatives. While megawatt-scale wind dynamos are expected to be economic when produced by the dozens or hundreds, quantity production by the thousands under competitive conditions is the key to making them attractive enough to command the capital investment they need. A government program to deploy them by the thousands, following the hydroelectric model, would surmount the hurdle to quantity production and thus pave the way for industry to enter the field in a big way on its own.

While wind power is the one of the solar-related energy sources that is technically and economically ready for large-scale deployment immediately, others will arrive at that stage in due course. What America needs is an agency whose mission is prompt deployment of energy sources as soon as they become ready for it, insofar as this may require government initiative. If established now, such an agency would start with wind power and be ready to expand into the ocean-thermal technology a few years later, as the most likely next candidate for prompt deployment.

The Decision and the Organization for Large-Scale Deployment

If the decision to undertake large-scale deployment of wind power could be made, the government already has the organizational framework within which to carry it out. But the nature of the organization has so far precluded making the decision. The trouble is that nuclear energy and solar-related energy have been like a fat little bird and a thin one in the same nest and the fat one has always been able to monopolize the nourishment and squeeze out the thin one. This was true in the days of AEC and ERDA when both undertakings were under the same administration and it is even more so in the next incarnation, the Department of Energy, where they are squeezed together in the same division under the same assistant secretary.

If the decision could nevertheless be made, the DOE type of organization is better suited than its predecessors to carry out such an undertaking. Its Division of Resource Applications could very appropriately be assigned the mission to deploy large-scale wind power soon rather than just to promote its deployment by others, and perhaps later to deploy other large-scale solar-related sources as they evolve past the R&D stage.

The grouping of nuclear and wind energies in the same division is part of the way the divisions of the DOE are differentiated according to managerial function rather than type of technology (DOE, 1977). At the assistant secretary level, there are fourteen divisions, fourteen boxes on the organization chart of which three are shown in figure 60. These three account for two-thirds of the budget and a fourth, Defense Programs, takes another one-fifth, leaving 14 percent for the other ten divisions.

The functions of the Energy Technology Division are further defined as follows. They include "research, development, and technology demonstrations in all energy areas—including solar, geothermal, fossil, and nuclear." Its activities "will focus primarily on making new energy technologies available for commercial (private or public) applications as early as possible . . . technologies grouped

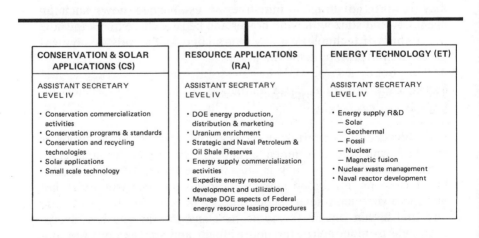

Fig. 60. Functions of the Department of Energy's three divisions directly concerned with solar and other power sources. A part of the DOE organization chart.

under this Assistant Secretary are still in the development stage. Once a specific project in a technology has been developed sufficiently for commercial demonstration the project will be transferred . . . to the Assistant Secretary for Resource Applications," or, in the case of solar heating and cooling, to the Secretary for Conservation and Solar Applications. "At the same time, further development of the base technology will continue under Energy Technology to improve efficiency, costs, etc."

With the 2-megawatt demonstration wind dynamo at Howard's Knob in North Carolina being managed by the local utility cooperative and with similar arrangements for two smaller machines, this outline of functions made it clear, at the time of the establishment of the Department of Energy, that megawatt-scale wind power was at the commercial demonstration stage and should be in the Resource Applications division. However, there appeared to be no intention of transferring it soon. Indeed, the Energy Technology division is essentially the former ERDA almost intact, except for such things as uranium enrichment and nuclear weapons.

Some of the functions of the Resource Applications division are thus defined: "The Assistant Secretary will begin Federal incentive programs to spur commercialization of technically developed energy technologies and will oversee the demonstration of those technologies (except conservation and solar heating and cooling) ready for commercialization."

It is significant that this division, besides having the mission to spur commercialization, is also in the hydroelectric power business itself. As the description of the newly organized DOE continues, "Also included under Resource Applications will be the four Interior Department power marketing administrations (Bonneville, Alaska, Southeastern, and Southwestern) and the marketing and transmission functions of the Bureau of Reclamation." As now interpreted, this means that the Department of the Interior still retains the function of building and operating hydroelectric installations, producing the electric power, while this division of DOE merely markets it. This mixed arrangement could be left as it is by Congress and used as an opportunity to bypass DOE and fund a large wind power deployment through this and perhaps other departments that might show more enthusiasm for it.

The description of functions also states, "The Assistant Secretary for Resource Applications will ensure that environmentally acceptable supplies of energy can be moved quickly into the commercial mainstream, and that institutional and other barriers to the introduction of new sources of energy supply are identified, and overcome." Two barriers to moving large-scale wind power into the commercial mainstream should be identified: (1) the capital-investment barrier since the principal present power sources, largely with past government promotion, have absorbed most of the available energy-investment capital and, (2) the quantity-production barrier that must be overcome by a substantial investment before costs are so manifestly competitive as to attract large amounts of capital. These barriers could be overcome by putting wind power alongside of hydroelectric power as a federal enterprise rather than just leaving it, as with nuclear power, for federal demonstration (not to mention other help) to "spur commercialization." Arranging the organization chart to facilitate doing this is one thing—that is done—but making the decision to do it is quite another and there seem to be "institutional barriers" impeding the decision in a government department so preponderantly interested in the currently exploited technologies.

The White House statement introducing the new DOE setup explained "while technology development is a continuum, different kinds of managerial skills are required along the way . . . the organization plan avoids entrusting all stages of a project to a single manager or organization." It is not clear whether this arrangement, requiring the transfer of responsibility for the progress of wind power, for example, from one division to another, will expedite or delay extensive commercialization. Entrusting progress to two managers may help in that the manager of the second stage might take the initiative and exert influence for transfer to his division for commercialization if the manager of the first stage tends to hang onto the project and keep it too long in the R&D stage. Indeed the Energy Technology division, being in effect ERDA renamed, inherits the AEC-ERDA tradition of keeping wind power moving along slowly in step with other less developed technologies. One possibility to break the deadlock would be for an aggressive assistant secretary for resource applications to claim jurisdiction and successfuly argue for the transfer with the top management of the DOE. He and his staff have a professional interest in doing so. He is, however, at a disadvantage in that he and the top management must rely for technical assessment of the situation largely on the technical expertise of the Energy Technology division. With its long interest in this technology and its competing interest in other technologies, this division might prefer protracted sequential demonstration before rapid commercialization.

The organization of the DOE effectively carries out President Carter's emphasis on achieving efficiency by consolidating governmental agencies and avoiding duplication. If motivations for change in its way of doing things can be effectively brought to bear in the new super-agency, this can be a very good thing. There is a danger, however, that it may be efficient only in doing things as they have been done at the expense of being inflexible toward new undertakings like really large-scale wind power. For people wishing to promote a rapid transition to large-scale wind power, a concerted approach should be made through the DOE organization, directing ideas and pressure to and through the assistant secretary for resource applications as well as the secretary of energy himself. If the president and his White House staff should want to do this, it should be a straightforward matter of directives and perhaps favorable appointments within the DOE and OMB as well as requests to Congress for specific appropriations. If a consortium of members of Congress or citizen and industrial organizations working through Congress should want to do it, the approach would include writing appropriate legislation into DOE appropriation bills. If there appears to be too much inflexibility in the DOE and this approach does not succeed rather soon, it might be wise to work instead on setting up a special agency to handle fuelless energy sources. It might be acknowledged that consolidation has gone too far, specifically in retaining the wind- and other solar-power development programs under the same management as nuclear power through all the reshuffling of agencies in their five-year history—from NSF to AEC to ERDA to DOE. The transition from AEC to ERDA was intended by some legislators to put solar power on an equal footing with nuclear power, but it failed to do so. Just one more reshuffling may be needed. With the administrator of an independent Fuelless Energy Agency preparing programs and going directly to Congress to solicit funds, the balance of appropriations between solar and nuclear efforts would be determined by Congress, not by the budget writers within the DOE. Such a reshuffling, however, seems not apt to happen soon.

Recommendations for Action

If either administrators or outside pressure groups decide to press for a more rapid introduction of large-scale wind power, action should be promoted both within the DOE and through other departments. Ideally, a broad-experience program testing several design concepts of large wind dynamos should be started as soon as possible. This lies definitely, however, in the province of the DOE. That department's lack of enthusiasm for speeding up large-scale wind power development may continue to delay progress in this direction. Particularly now that a megawatt-scale wind dynamo of an alternative design is being demonstrated in the private sector (see fig. 57) it would be highly desirable to start substantial federal deployment very soon without waiting for even the start of a broad-experience demonstration program. Both the Department of the Interior and the Department of Agriculture would probably be interested in undertaking such a mission. The Department of the Interior, whose full name is Department of the Interior and Insular Affairs, could appropriately deploy hundreds of megawatt-scale wind dynamos to supplement its hydroelectric generating facilities in the western states and also could install wind power in remote islands that have good winds, yet import oil over long distances. The Department of Agriculture could install large wind turbines for both large-area and high-pressure irrigation.

While these possibilities should be promoted for the sake of a quick start, acceleration and broadening of the DOE program should also be encouraged. Its broader demonstration program would improve the long-term prospects and its deployment program could be directed to end uses not covered by other departments' activities.

If we consider separately the agenda within the DOE, there is a wide range of options that might be chosen as the objective of the initial promotion, ranging from a modest change to a drastic change from the present rather narrow program. Each of them involves at least abandoning the directive that demonstrations of large wind dynamos must be sequential and instead proposes gaining broad experience soon from a number of parallel demonstrations of different machines, as has already been discussed. Some of these options that may be recommended are as follows:

1. *A minimal broad-experience program.* The present megawatt-scale demonstration program involving about ten million dollars in two years, or twenty million in three or four years, should be stepped up to about fifty million dollars in the next two years. The specific demonstrations in addition to those already planned (the 2-megawatt, 200-foot diameter Howard's Knob machine and the similar 300-foot diameter machine to follow) should include one floating offshore machine (similar to either figures 29 or 31, but perhaps somewhat smaller), the British tripod design of figure 12 and at least one land-based three-bladed design, perhaps exploring the economies of using guy wires as in figures 8 or 10. A duplicate of each, less costly than the first, should preferably be included. Whatever the list of designs chosen for testing, it is imporant that

full-scale experience with several promising designs should be gained to determine which are best for production.

2. *A large broad-experience program.* This is the broad-experience program discussed in chapter 11, involving an expenditure of one hundred fifty or two hundred million dollars in two or three years. The land-based machines to be demonstrated could include, for example, a duplicate of the 300-foot diameter turbine on a 500-foot guyed tower, the aforementioned tripod and three-bladed machines, a machine using the leaning tower concept, a machine seeking on a megawatt-scale some of the economies suggested by the Gedser and Cuttyhunk turbines, and, perhaps also on a megawatt scale, whichever of the vertical-axis turbines appears most promising at the time. There should also be a bottom-mounted, shoal-water offshore machine and about three floating offshore types, one of them a single-turbine on a buoy flotation, another with two or three turbines similar to figure 29 and a third with multiple turbines on a platform mounting perhaps similar to figure 31.

In such a program, variety of types is more important than numbers of machines tested and judgment must be exercised in promoting it to avoid failing to achieve any program by seeking too much. If it seems politically achievable it would be desirable also to have the demonstration and testing program include not just one but several machines of each type. This takes advantage of the fact that the cost of the first machine includes the engineering costs so each of the duplicates costs perhaps only one-third as much but is valuable for acquiring operating experience soon. Several of the land-based types with the duplicates should be deployed in an array on the western Great Plains.

3. *The minimal broad-experience program plus federal deployment.* In the interest of speedy massive deployment, the decision would be made in this plan to start the minimal broad-experience program and federal deployment at about the same time, basing the initial deployment on experience with the Howard's Knob 2-megawatt wind dynamo without waiting for the results of the broader demonstration and testing program. A large array in the western Great Plains would be built up gradually over perhaps a ten-year period at a rate of perhaps a hundred or more of the machines a year, those of the first two or three years being of the original model but with the possibility of shifting to installing other models at an accelerated rate as the result of later experience. The experience and the ongoing production would be expected to cata-

lyze commercially financed installations in other arrays and localities.

4. *The large broad-experience program likewise plus federal deployment*. Federal deployment of a land-based array would start at an
early date as above. The broader experience would provide a firmer
basis for expecting improvement in later models, with a wider selection to choose from. Here the broader variety of experience at sea
would provide a firmer basis for starting federal deployment of an
offshore array as well.

Sources of Pressure for Action
With alternative power so sorely needed and wind power so ready to
supply it, the decision for some such option should come from
within the administration. The past record, however, makes this
seem unlikely. If it does not, pressure exerted on and by Congress
might accomplish it. Unlike some other power sources, large-scale
wind power has no one making money from it and able to afford
expensive lobbying in its favor. No labor unions represent present
jobs in it that must be preserved. Instead, the pressures must be
based on future expectations that can be generated only by some
intellectual exercise and imagination and persuasion. Deployment of
large-scale wind power will mean many jobs all over the country for
people in the auto industry making gears and shafts, those in the
aircraft industry making blades, in the structural steel industry making towers, in the electric industry making generators, and eventually in shipyards making flotation, among others. Impressing enough
members of Congress with the prospect for jobs in their constituen-
cies, not only in the states where the big winds blow, but all over the
country, may be the key to getting the decision made. Decisions for
hydroelectric and rivers and harbors projects have traditionally been
made by congressional "logrolling" and the same might apply to
wind power.

    As has been mentioned, the AEC-ERDA-DOE federal energy
program has had a policy not to carry out parallel demonstrations
of large units in the same general field, but rather to keep such
demonstrations sequential. For technologies requiring very large,
several-hundred-million-dollar, central-station type generating units
to have some prospect of being economic, such as ocean-thermal
and solar-thermal plants, this seems like a fairly reasonable limitation. With the present pressing need for alternative energy sources
soon, it does not make sense as applied to wind power in which the
individual generating units cost around five or ten million dollars

for the first demonstration and are expected to be little more than one million dollars apiece in serial production. What is immediately needed is a decision, which could and should be made within the DOE with the concurrence of OMB, that this insistence on sequential demonstration does not apply to megawatt-scale wind dynamos. A broad-experience program would be a natural consequence of such a decision.

With all the urgent uses there are for wind power, with or without storage, it seems a form of national insanity that we do not get started soon on a federal program somewhere within this range of options. If a decision cannot be made promptly on more, it should at least be made on a minimal broadened experience like program 1 as a starter, leaving until later as the results come in the decision to expand it and supplement it with a federal deployment project as in programs 3 and 4. But that would be a paltry reaction to the opportunity. To do less than adopt an option as forthright as program 4 would under present circumstances seem remiss.

Appendix 5

Later Developments in Wind Power

During the years 1978–80 the long approach to the megawatt scale in the federal wind-energy program finally came to fruition with the completion of MOD 1, the General Electric 2 MW wind turbine at Howard's Knob near Boone, North Carolina, and in the Columbia River Gorge in Washington, the near-completion in triplicate of MOD 2, the fine Boeing 2.5 megawatt machine with a 300 foot blade span. MOD 1 has been a disappointment, encountering several troubles and costs too high to hold promise for commercialization. Some of the trouble is apparently related to having eliminated the flapping freedom provided by the hinges near the hub of the Smith-Putnam machine shown in figure 6.

MOD 2 is very differently designed and is much more promising. Most of the flexibility of the hinges in the Smith-Putnam design is achieved more simply and effectively by using the teeter principle introduced by Hütter in his wind dynamo shown in figure 10. The two blades are in a straight line like a stiff beam but have a common hinge permitting a teetering change of angle with the shaft. Teetering is also used in helicopters and the availability of computer codes for that purpose was helpful in the design of MOD 2, a nice example of applying modern technology to an old art. Centrifugal and Coriolis forces play an important part in keeping the blade circle nearly, but not quite, normal to the shaft during yaw as the wind shifts, thus avoiding passing on vibrations to the shaft. A three-bladed rotor is another way to avoid these vibrations, so there is a trade-off between the cost of a third blade and the cost of the rather elaborate rubber-metal-sandwich teeter bearing. One economy in MOD 2 is the provision of variable pitch for only the outer 30 percent of the blades, but it otherwise resembles the picture in figure 56. The slender tower is known as a "soft" tower, its principal vibration frequency being lower than the frequency of the passing blades.

The design of MOD 2 was commenced well before the completion of MOD 1. This represents an encouraging but small relaxation of the old sequential policy that permitted only one large-scale development at a time and leaves hope that a greater breadth of large-scale experience may be sought in the future.

As an important step towards the building of larger wind farms, the initial authorization is for not one but three MOD 2s. They are in the Columbia River Gorge, about a hundred miles east of Portland, where there are good winds and are grouped to provide a first test of wind interference between adjacent large turbines.

The plan for federal wind-power development proposed in 1976, as charted in figure 47, envisaged two megawatt-scale experimental units built in 1978. Instead there was one in 1979 and three of the second and better model in 1980–81. These three, with a total of 7.5 megawatts, about fulfill the expectation of a multiunit demonstration involving a 10 MW pilot plant as anticipated for 1980 in the chart of figure 47. There a 100 MW demonstration that will take longer is predicted for 1981. That chart mysteriously postpones expectation of production facilities until the next century. Boeing is proposing instead, as soon as it gets enough orders, to convert a plant for assembly line production of twenty MOD 2s a month and expects that the hundreth unit would cost two and a half million 1980 dollars, or $1,000 per kilowatt. In terms of 1977 dollars, this means about $700 per kilowatt of installed capacity or $2,000 per average kilowatt, considerably more than the expectations based on table 2, but still considerably less than the $2,500 per average kilowatt estimate for future nuclear power plants, including fuel, explained in chapter 9. More recent nuclear experience seems to bear out that estimate.

The 200-foot tower height of MOD 2 was chosen arbitrarily to provide convenient ground clearance. Calculations by Boeing engineers of the increased power from the stronger winds with increased tower height showed a nearly constant cost effectiveness in the range 175 to 250 feet for a conventional wind-shear pattern in which the speed increases by about 12 percent as the height is doubled (expressed as eα with $\alpha = 1/6$). In some locations the increase is greater than that, such as about 20 percent on doubling (or $\alpha = 1/4$) observed at the Plum Brook, Ohio, wind turbine site. (An increase of 20 percent means 73 percent more power, with power proportional to speed cubed.) From this it seems clear that in some regions where the surface winds are not particularly strong, it will pay to build wind turbines with towers 500 feet high or more, an idea promoted by Percy Thomas in an earlier federal program (fig. 7), but arbitrarily rejected in the modern program (Wahrenbrock, 1979). A start is being made to measure higher altitude winds.

The successful development of MOD 2 not only justifies putting it in quantity production but it can have an impact beyond the deployment of this particular model. By having built confidence in wind power as an important energy option, it can be the basis for starting very large federal wind power projects. If appropriately administered to encourage competition, these could, as a market, stimulate a healthy variety of independent developments.

Rejection of Offshore Wind Power

No such progress has been made towards developing offshore wind power. On the contrary, a strangely evasive study report has been produced that seems contrived to make offshore wind power look unattractive, and the report has been used in Washington to try to justify the longstanding decision not to pursue this important option in the federal program. The origin of the Westinghouse Offshore Windpower Feasibility Study and its possible implications are outlined in chapter 11 and appendix 4, and in a further reference (Inglis, 1979). The study was mandated through legislation by Congress in 1975 and the report on it (Kilar, 1979) appeared in draft form almost two years behind schedule in mid 1979. On the question of cost effectiveness that has been considered decisive by DOE, the conclusion is negative, but on this question the report is remarkable for its transparent irrelevance. It sets up a straw man and knocks it down. The designs of offshore floating wind power systems considered are selected with no regard for efficient and economical seaworthy design and construction. According to the project manager, the main emphasis in the study was to be on reliability of the cost estimate, which meant staying within the "state of the art." The state of the art consisted of designs of land-based windmills, so the procedure was to design a platform at sea simulating a site on land on which to erect an already-designed, land-based wind turbine modified to withstand corrosion. In the words of the report (vol. 3, pp. 2–21) "As wind energy conversion moves from land into the offshore environment, it becomes necessary to construct the 'real estate' upon which the wind turbine generators will be mounted." The floating platforms designed were made very large to avoid tipping over and thus expensive enough that only the very largest wind turbines were considered, at least as large as MOD 2 though with the broad-based tower of MOD 1. The ultimate absurdity from the point of view of cost effectiveness is the fact that each platform is moored with many anchors so that it, like a land site, will not rotate. The usual yaw mechanism aloft, as needed on land but a superfluous expense at sea, then serves for facing into the wind. Providing and maintaining the many moorings is a substantial part of the projected cost. It was taken for granted in previously proposed designs, such as those sketched in chapter 4 of this book, that a single mooring would be used so that riding on the mooring would provide the orientation into the wind and no further yaw mechanism would be needed. This also leaves the attractive option of mounting multiple turbines on a single flotation. This is one economical feature of offshore wind power to help compensate for the extra cost of flotation and mooring as compared with land siting. Another is the possibility of localized assembly under shipyard conditions, also mentioned in chapter 4. Neither of these economies is exploited in the Westinghouse study. One remarkable feature of the draft report is that, while it makes ample references on other matters, it makes no mention at all of previous literature on the design of offshore wind power systems, thus obviously avoiding consideration of possibly much better designs.

Two floating platform designs were considered in detail. In one the "real estate" is a 100-foot square slab perched 20 feet above mean sea level atop a 30-foot-diameter vertical pipe floating as a spar buoy. It is moored by eight cables radiating like the eight legs of a spider, the weight of each being carried by an outboard float. The other is a four-cornered raftlike structure moored by four vertical chains under tension to keep it level despite the sea. These presumably must remain taut even in the trough of a storm wave to avoid jerking, requiring excess buoyancy and anchor weight pulling against each other.

In other matters the report has positive features, such as undersea electric cable data. And most importantly, it does certify the technical feasibility of floating offshore wind dynamos. This should prove useful as a basis for a political decision to award a contract for design, construction, and demonstration of floating offshore wind power, without requiring further feasibility study, to some organization that wants to make a technical and economic success of it.

At an internationally attended meeting on wind power (the fourth biennial workshop at Washington in late 1979), a DOE representative presented this deficient report as proving that offshore wind power is much too expensive and concluded that it would not be pursued in the United States' program. Thereupon representatives of several other countries rejected the validity of the report and said they intended to go ahead with offshore wind power anyway. Apparently being faced with the prospective embarrassment of being left behind, the DOE people then joined with those from Sweden, the Netherlands, and Great Britain in supporting an international investigation to assess the practicality of offshore windpower. Since each of these other countries seems genuinely interested in making a go of it, this new approach is apt to come up with a more favorable result, perhaps despite American skepticism. If it does, it is to be hoped that this will encourage some nongovernmental group to undertake a demonstration without awaiting further governmental initiative that might be slow to materialize either here or abroad.

Government Policy

Growing emphasis on wind power in Congress has been providing increased funding, sometimes more than sought by DOE, and has encouraged broadening of federal wind-power effort. DOE policy, however, has hewn closely to the path charted earlier, confining the government role to support of development and demonstration while leaving the rest of commercialization to private enterprise. This has meant selecting specific projects for ample support, including few large machines and rather many small ones. There has been a widespread feeling among those concerned that more rapid development of practical machines could be obtained with the same funding by subsidizing the private purchase of wind turbines, plus government establishment of wind farms, to create a market, stimulating private investment and spontaneous effort.

Demonstration of smaller wind dynamos is organized at Rocky Flats, Colorado, where about a dozen models are being tested, most of them having been developed under contract. One 8 KW model, for example, has received over half a million dollars of federal funding. An impressive number of small firms and independent individuals have expanded their developing and marketing of small wind turbines in this period.

The high level of funding of specific projects is partially justified by the need for sophisticated modern engineering to assure long life, including a lot of vibration analysis and fatigue testing. However, some of the high cost of both development and prospective sales goes to supporting the corporate superstructure of the large firms, some of them aerospace-oriented, whose competence is considered well-established. Sophisticated modern engineering is also available to less prestigeous firms and they may later succeed in bringing prices down on big machines as is happening on small ones, if they can overcome the head start provided to large competitors by federal support.

Legislation

The Wind Energy Systems Act of 1980 directs the Department of Energy to take new initiatives both to accelarate the construction of large wind turbines and to broaden the scope of the program. It sets a goal of 800 megawatts by 1988 from a total of 387 large (mostly megawatt-scale) windmills partly or fully financed by the federal government, to be installed at a rate increasing from 10 large turbines a year in 1982 to 162 per year in 1988. The projected cost of the eight-year program is about $100 million, of which $65 million is for large-scale prototypes. To the previous government role of supporting research, development, and demonstration, this legislation adds the new role called "technological applications," meaning installation of wind energy systems. Besides direct government purchase, both grants and loans covering part of the costs are to be made to outside organizations. The DOE is also to supply funds for large wind systems to the Water and Power Resources Service and the federal power marketing agencies such as Bonneville. In a longer perspective, the bill mentions a goal of 1.7 quads per year (or 57 average gigawatts) from wind by the end of the century, but this would seem to justify a more vigorous start now.

By way of encouraging the independent initiatives that might provide some of the broad experience recommended in chapter 11 and in this appendix, the DOE is authorized to make grants for demonstrating "a variety of prototypes of advanced wind energy systems under a variety of circumstances and conditions." The bill as passed by the Senate further specified "on land and offshore," but this was deleted in Senate-House conference in conformity with DOE opposition to going offshore.

This 1980 legislation was initiated for the previous year but was delayed by the long congressional hastle over synthetic fuels, with which wind energy was included in a general energy bill. The "synfuel" program is given such overriding priority that a special governmental corporation for it is established, making it independent of DOE. Its initial expenditures are anticipated at around $20 *billion* with a total of $88 billion proposed for some years later. This is about 100 times the new proposal for wind power. Being promoted by strong vested interests, the synfuels program is receiving such massive public backing despite the expectation that its product will cost several times as much as the fuel it replaces, if the innovative development succeeds at all. Wind power, lacking such powerful promotion, must apparently undersell the competition substantially before attracting that kind of attention. What it could do with it!

In the Private Sector

A Connecticut firm, Hamilton Standard, produced a design that was rejected by DOE for the blades of MOD 1. Thereupon the firm teamed up with a Swedish shipbuilding firm, Karlskronarvet, to produce large two-bladed windmills based on Swedish experience. The first of them rated at 3 megawatts will be part of the Swedish program and the second, rated at 4 megawatts, is being bought by the Water and Power Resources Service (formerly Bureau of Reclamation) of the United States Department of the Interior. The blade diameter is 255 feet (less than MOD 2's) and tower height about the same, 260 feet (greater than MOD 2's). It is to be installed in 1981 near Medicine Bow, Wyoming, where the Department of Interior's Water and Power Resources Service (formerly Bureau of Reclamation) has for some time been preparing wind-power sites. This marks the entry into the megawatt-scale field of another serious competitor not supported by DOE. (It undersold Boeing with a bid of about $6 million for the 4 megawatt windmill). Even more importantly, it also marks the entrance of another government department into large-scale wind-power deployment, appropriately as a hydroelectric supplement. It opens the prospect of government wind farms establishing a competitive market and encouraging diversity in keeping with the intent of the new legislation.

Among large wind dynamos, the most impressive one developed in America that has not had federal assistance is the Schachle rotating-tripod design cited in appendix 4, the prototype of which is shown in fig. 57 and the frontispiece. The 3-megawatt machine for Southern California Edison Company was completed in 1980 after there was a two year delay because the small Schachle firm that initiated it could not obtain adequate funding. Completion of the project as undertaken by the Bendix Corporation (with Mr. Schachle as director of research) has involved superposition of corporate organization and some redesign that have increased the cost well beyond what was anticipated, making it approach that of MOD 2. It will be of great interest to see how this compares with other large machines as testing proceeds.

One small firm that is moving up into the megawatt range from 250 kilowatt turbines and smaller, of which several have been sold, is the Energy Development Company (Merhkam) of Hamburg, Pennsylvania. Their 2-megawatt unit has a cylindrical tower like that of MOD 2 and a rotor having six slender metal blades with fixed pitch and no taper. The many blades permit slower rotation that reduces vibrations. The per-kilowatt cost is about one-third of that expected for quantity-produced MOD 2s.

In the range under a megawatt, one line of vertical-axis wind turbines (VAWTs, they are called) reached the stage of commercial production but unfortunately came to a bad end. Building on its experience making blades for the development of Darrieus turbines at Sandia Laboratories in New Mexico (see fig. 20), the aluminum company Alcoa started marketing Darrieus dynamos in sizes up to half a megawatt at a relatively low price. An important economical feature was that the hollow blades could be made by the simple process of extruding aluminum. However, the program came to grief and was abandoned when several of the windmills shook themselves apart and fell to the ground, creating grave doubt of the viability of at least this vertical-axis approach.

Development Abroad

The DOE program of large-scale wind turbines has been criticized for its lack of breadth, concentrating as it does on two-bladed turbines following the Smith-Putnam precedent. It has by now investigated a variety of hub, blade, and tower designs within this limitation, and that may in the end prove to have been the right approach. The structural simplicity of a two-bladed rotor has its obvious advantages and it may be that its disadvantages can be overcome economically with continued engineering development. One cannot be sure of this in advance and the need for mounting a broad-experience program before freezing on any one design for large-scale wind farms has been emphasized in chapter 11 and appendix 4. Such a program has not been initiated in the United States effort but some breadth of experience is about to be achieved worldwide through the diversity of several other important national programs, particularly those in Germany, Denmark, and Sweden. Further diversity of construction and financing methods is to be found among the private enterprises not supported by DOE in the United States. Between them, these approaches will give experience with large wind turbines having one, two, three, and even up to six blades, with a variety of tower types and heights. Unfortunately, none of them as yet contemplated is to be floating offshore or more than 360 feet in tower height.

Of the national large-scale wind energy programs, that of the West Germans seems to be the most imaginative and innovative, but completion of the very large machines being planned is not expected until the period 1982–85. The first, known as GROWIAN I, is to be quite conventional, rather similar to MOD 2 in size and general characteristics: a two-bladed, downwind, slender tower, but with a different hub structure and taller tower. It is rated at 3 MW in a 26 mile-per-hour wind. The second, GROWIAN II, is to be much larger with a single blade (!), following an early experimental demonstration by Professor Hütter. The single blade is counterbalanced and teeters, being held out almost normal to the axis against the wind pressure by centrifugal force. There is a prototype in nature in a maple seed that slows its fall and rides on the wind by flying like a helicopter (or autogyro) with a single wing counterbalanced by the seed itself. GROWIAN II will be rated at 5 MW and have a blade circle 475 feet in diameter on a tower 360 feet high. The concept is first to be tested with a one-third scale model. The third large wind-power project is even more imaginative. It is more a matter of direct solar energy than natural wind energy, using solar heat to create an artificial vertical wind to drive a turbine. The solar collector is a very large circular roof just above the ground connecting to a cylindrical tower at the center up which the air heated under the roof rises to drive a wind turbine, a thermal-upwind power plant. The units being considered are huge 10 to 100 megawatts. The 100 megawatt one would be 3 kilometers in diameter.

The Swedish megawatt-scale program is similar to the American in being confined to two-bladed turbines. For the sake of direct comparison of various engineering details, two large windmills with about the same dimensions are being constructed in parallel and due for completion in 1981. The two machines are similar in having about a 75-meter (250-foot) blade span and cylindrical towers 80 meters (260 feet) high but they purposely differ in many respects: construction by different pairs of firms, rated power of 2 or 3 megawatts in winds of 28 or 32 miles per hour, "soft" steel or rigid concrete towers, downwind or upwind rotors with all-glass-epoxy or partly steel blades, even different gear boxes and generators. While well-planned for assessing these design options the program seems to be progressing no more rapidly than the American program and is indeed about a year behind it in attaining the megawatt scale. The costs to the government for the large demonstrations are about half as great.

The Danish program concentrates on three-bladed rotors, following experience with the Gedser mill (fig. 8) and others. Two wind dynamos of identical dimensions have been built, differing only in control mechanisms and rotor design, again for the sake of testing the viability of variations in these features by direct comparison of the two. Each machine is rated at 630 kilowatts in a 29 miles-per-hour wind, near enough to a megawatt to provide an important variety of experience applicable to megawatt-scale turbines. One of them has a rotor braced with stays, as in the old Gedser mill, but with the stays extending out only one-third of the radius. Beyond that the pitch of the blades is somewhat variable, permitting stalling for overspeed control. Small wind dynamos in the range 10 to 55 kilowatts for connection to utility grids are being demonstrated and manufactured in Denmark by about a dozen different firms. Most of the models are three-bladed, horizontal-axis machines, four of them with fantails. Most of them are being tested at a government test facility similar to that at Rocky Flats in the United States.

Conclusion

This book has emphasized that wind power is an energy resource whose time has come, requiring no technical breakthrough but a rather drastic change of priorities and financial commitment to instigate the scale of engineering and industrial effort that could make it meet a substantial part of our energy requirements in the next few years. Despite the growing need for alternative energy sources, these last two years of the decade have seen no such drastic changes but rather a continuation of the business-as-usual attitude. There have been important advances both here and abroad, but only at a pace that may bring wind power into its own towards the end of the century. It has been suggested here that a high-priority program should start with a couple of years of an intensive broad-experience phase, with design and testing of a variety of quite different models as a basis for choosing those that will probably be most cost-effective in quantity production. On a worldwide basis, some of that is being achieved in these years, bringing us close to the point where a large committment should be made to extensive deployment in large wind farms, starting as a hydroelectric supplement but with a view to going much further. A small beginning has been made for organizing a federal project outside the DOE.

The passing years are remarkable for the opportunities that have been missed. The business-as-usual attitude has left important innovations unexplored. These include offshore wind power to exploit the strong ocean winds that has been studiously rejected, higher-tower wind power such as promoted by Percy Thomas (see chap. 8 and fig. 7) to exploit the steadier winds aloft (Wahrenbrock, 1979), and the concept of an integrated community energy system combining wind power with direct solar power and interseasonal thermal storage (Engelke, 1978). It is noteworthy that European designers have found it economical to use considerably higher towers than those of the large American windmills, a step towards really high-tower windmills. Now that tens of billions of dollars are being turned toward synthetic fuel, for example, we should be thinking what could be done with such amounts, or simply with one billion dollars a year, devoted to big windmills. Just for example, half of the billion (1980 dollars) could initially go to buying two hundred MOD 2s a year, the output of the factory that has been proposed to produce twenty a month, then perhaps a quarter of a billion dollars for the incidental expenses, including transmission lines and some storage, to use them, leaving a quarter of a billion dollars for purchases from competitive suppliers probably at considerably lower prices. The MOD 2s alone would in two or three years be equivalent to a large gigawatt nuclear plant and, despite their high initial price, at roughly the same cost. An initiative on such a scale, as a starter, would represent an appropriate priority.

There has been much talk in recent years about the need to develop alternative power sources and about the need to relieve unemployment. There has been too little action. A large auto manufacturer has been rescued by federal financial assistance to preserve jobs making more autos to consume scarce fuel, an example of the political pressures to perpetuate old ways rather than to start something new. It remains urgent that fewer workers and less industrial capacity should be devoted to making big autos and more to making big windmills as one important part of developing the diversity of energy sources we need.

Bibliography

Abelson, P. H. "Energy and Climate." *Science* 197, 941, September 2, 1977.

Anderson, B. "The Sun in a Drawer." *Environment* 17, 8, October, 1975, p. 36.

Andrews, J. W."Energy-Storage Requirements Reduced in Coupled Wind-Solar Generating Systems." *Solar Energy* 18, 73, 1976.

American Physical Society (APS). "Report to the American Physical Society by the Study Group on Light Water Reactor Safety." *Reviews of Modern Physics* 47, supp. 1, Summer, 1975.

Atwood, G. "Strip Mining of Western Coal." *Sci. Amer.* 233, 6, December, 1975, p. 23.

Barnes, J. "Geothermal Power." *Sci. Amer.* 226, 70, January, 1972.

Barney, G. O., ed. "The Unfinished Agenda." (A report sponsored by the Rockefeller Brothers Fund). New York: T.Y. Crowell, 1977.

> Policy proposals on many environmental issues. Proposes a national Planning Board with long tenure and long-view advocacy function to guide the usual short-range planning by industry and elected officials. On energy, too little consideration of scale of need and of technical and economic readiness of particular renewable sources to meet it soon. Mentions using the wind to drive only small pumping windmills. Emphasis on appropriate energy locally generated and on avoiding electrification neglects opportunity to use electric transmission from where the best wind is to where there is the biggest need.

Beedell, S. *Windmills*. New York: Charles Scribner's Sons, 1976.

> A charming pictorial exposition of the craft of millwrights of old and including an old wooden windmill in Denmark in commercial operation up to 1973.

Berger, J.; Reis, M.; and Rudolph, R. "Inside Carter's Energy Plan." *Not Man Apart* 12, 7, June, 1977, p. 8.

Bethe, H. A. Oversight Hearings, *Subcommittee on Energy and Environment*. House Committee on Interior and Insular Affairs, Washington, April 28, 1975.

———. "The Necessity of Fission Power." *Sci. Amer.* 234, 1, January, 1976, p. 21.

> Solar-nuclear cost comparison neglects plant factor of nuclear and assumes solar construction would take as long and have as great interest and escalation costs during construction. Accepts Rasmus-

sen results with minor modifications and cites, as check on those estimates from experience, the fact that "there has never been a loss of coolant in 300 reactor years. . . . " Dismisses wind power as presenting "its own special difficulties."

Betz, A. "Das Maximum der theoretisch möglichen Ausnutzung des Windes durch Windmotoren." *Zeitschrift für das gesampte Turbinenwesen* 17, 20 September 1920.

Boer, K. W. "Solar Heating and Cooling of Buildings." In *Proceedings of the First Symposium on RANN,* edited by J. Holmes. Washington: National Science Foundation, 1973, p. 50.

Brown, L. *Ecology and National Security.* Report of Worldwatch Institute, 1977. Reviewed in *Science* 198, 712, November 18, 1977.

Saving oil and other reserves with alternative energy sources may be more important for national security than new weapons systems that command so much money and effort.

Browning, J. "Hydraulic Wind Turbine Patented." *Wind Power Digest* 9, 21, Summer, 1977.

Brulle, R. V. "Feasibility Investigation of the Giromill." Sandia Laboratories Workshop Report. SAND 76-5586, July, 1976.

Bulbransen, E. A. *Bul. Atomic Sci.* June, 1975, p. 5.

Bullard, E. C. Proc. Roy. Soc. (London) A197, 433, and A199, 413, 1949.

Bupp, I. C., and Dinsimoni, M.P. "The Breeder Reactor in the U.S.: A New Economic Analysis." *Technology Review* 76, 26, July–August, 1974.

Bupp, I. C.; Cerian, J. C.; Dinsimoni, M. P.; and Trietel, R. "The Economics of Nuclear Power," *Technology Review* 77, 14, February, 1975.

Bureau of Economic Analysis (BEA). "Purchasing Power of the Dollar as Measured by Wholesale Prices, Business Statistics." In *The Biennial Supplement to the Survey of Current Business, Bureau of Economic Analysis, U.S. Department of Commerce,* 1975, p. 50. Supplemented by *Survey of Current Business,* 56, 11, 1976, p. S-9.

The inflation factor from 1945 to 1977 (extrapolated slightly from October, 1976) is 3.6. From 1972 to 1977 it is 1.65, for example, it rose by a fairly steady 3 percent per year from 1967 to 1972 and then leapt to about 13 percent per year for the next three years.

Burke, K. C., and Wilson, J. T. "Hot Spots on the Earth's Surface." *Sci. Amer.* 235, August, 1976, p. 46.

Business Week (BW). "Why Atomic Power Dims Today." *Business Week,* November 17, 1975, p. 98.

Calef, C. E. "Not Out of the Woods." *Environment* 18, 7, September, 1976, p. 17.

Northern-type forests more than area of United States required to produce our energy use. Sugarcane, area of Texas, 50 times water of Colorado River, 7 million tons potash, plus other fertilizers.

Carey, W. D. "Last Resorts." *Science* 197, 329, July 22, 1977.

Carter, J. "VAWT" (vertical axis wind turbine) Research at Sandia Labs.

Wind Power Digest, 9, 42, Summer, 1977.

> Tests on 5-meter Darrieus rotor show it about half as effective for a given projected area, as a propeller-type turbine.

Carter, L. J. "Failure Seen for Big-Scale, High-Technology Energy Plans." *Science* 125, 764, February 25, 1977.

> Similar to Lovins (1976).

———. "Nuclear Partners, Adversity Breeds Trouble Between Dow and Utility." *Science* 196, 162, March 14, 1977.

———. "Nuclear Wastes: Popular Antipathy Narrows Search for Disposal Sites." *Science* 197, 1265, September 23, 1977.

Clarke, A. C. *Profiles of the Future.* New York: Harper and Row, 1973.

Claude, G. "Power from the Tropical Sea." *Mech. Engineering* 52, 12, December, 1930.

Clews, H. M. "Wind Power Systems for Individual Applications." *Wind Energy Workshop Proceedings,* December, 1973, p. 164.

Cochran, T. B. *The Liquid Metal Fast Breeder Reactor.* Baltimore: Johns Hopkins University Press, 1974.

Cochran, T. B.; Speth, J. G.; and Tamplin, A. R. "A Poor Buy." *Environment* 1, June, 1975, p. 12.

> LMFBR will be more than $100 per installed kilowatt more than LWR. AEC prediction is that electric power demand in year 2020 will be 15 times that in 1975. Assumes learning curve for LMFBR only, not LWRs will bring breeder down to LWR capital costs by year 2000, having started in 1987. Of that 15 times, NSF panel estimates 20 percent can come from solar, AEC estimated 6 percent from geothermal even without hot rock technology. If hot rock technology works, could make 400 gigawatts by year 2000.

Comey, D. D. "Will Idle Capacity Kill Nuclear Power?" *Bul. Atomic Sci.* 30, November, 1974, p. 23; 31, February, 1975, p. 40; 31, October, 1975, pp. 40, 42; and commentaries by Chase, J., 31, February, 1975, p. 30; Morgan P., and Lindhe, S., 31, October, 1975, p. 38; Netachert, B. C., 31, October, 1975, p. 42.

Commoner, B. *The Closing Circle: Nature, Man and Technology.* New York: Knopf, 1971.

———. *The Poverty of Power: Energy and the Economic Crisis.* New York: Knopf, 1976.

Commoner, B.; Boksenbaum, H.; and Corr, M., eds. *Energy and Human Welfare, The Social Cost of Power Production.* New York: Macmillan Information, 1975.

Coste, W. H., and Lotker, M. "Evaluating a Combined Wind Power/Energy Storage System." *Power Engineering,* May, 1977, p. 48.

> Concludes that, with average wind 15 mph, wind power with storage in New England is cheaper to the utility company than nuclear or coal power if used for peaking periods lasting up to 30 percent of the time.

Coty, U., and Vaughn, L. "Effects of Initial Production Quantity and Incentives on the Cost of Wind Energy." Lockheed California Company. Paper delivered at the AAAS Meeting, February 20–27, 1977, Denver, Colorado.

> Treats cost of 2-megawatt wind dynamo as 30 percent cheaper in production run of 1,000 than in run of 100. Finds wind power economic as fuel saver alone in 7-meter-per-second wind regime with production run of about 300 units. Notes that quantity production is needed to reduce unit costs enough to induce utilities to buy in quantity, a "chicken and egg" cycle that must be broken somehow.

Crawley, G. M. *Energy*. New York: Macmillan, 1975.

> Good general text. The obsolete cost estimate favoring nuclear over coal on p. 117 ignores the trend of figure 45. The assessment of wind power potential on p. 185 presents a particular proposed development as the maximum possible.

Daniels, F. *Direct Use of the Sun's Energy*. New York: Ballantine, 1964.

———. "Utilization of Solar Energy—Progress Report." In *Proceedings of the American Philosophical Society* 115, 90, 1971.

d'Arsenval, J. *Revue Scientifique,* September 17, 1881.

DOE. *Organization and Functions Fact Book*. Washington, D.C.: Department of Energy, 1977.

Donovan, P., et al. *An Assessment of Solar Energy as a National Energy Resource*. Report of the NSF/NASA Solar Energy Panel, University of Maryland, December, 1972.

> This preceded the similar report (RANN, 1973) and the withheld Panel IX Report. Purports to give maximum energy production from wind power from the Great Plains, for example, as 24,000 average megawatts. However, this is calculated with extremely sparse coverage of a large area (390,000 square miles), with an average (though they are clustered) of one 2-megawatt wind dynamo per 22 square miles (and a high capacity factor of 69 percent)! Sees total potential United States production including offshore and Great Lakes as 175,000 average megawatts. Recommends $0.6 billion, ten-year wind energy development and deployment program including several 100-megawatt installations. This report was available during preparation of the national energy plan (Ray, 1973) but was ignored, the planned expenditures for wind power being only 5 percent of those recommended in this report. This is one bit of evidence suggesting that wind power has been deliberately supressed.

Eldridge, F. R. "A Preliminary Federal Commercialization Plan for WECS." *Amer. Wind Energy Ass'n Newsletter,* Spring, 1977, p. 14.

> Suggests Federal WECS Corporation to initiate mass production by stockpiling wind turbines and parts for commercial sale. Southwest Project of eleven utility companies considering combining wind power with northwestern hydro power.

Eldridge, F. R., ed. "Second Workshop on Wind Energy Conversion Systems." Washington, 1975, report NSF-RA-N-75-050. McLean, Virginia: The Mitre Corp., 1975.

Energy Research and Development Administration (ERDA). "A National Plan for Energy Research, Development and Demonstration." Vol. 1, The Plan. Vol. 2, Program Implementation. Report ERDA 76-1, 1976.

Engelke, C. E. "A Self-Contained Community Energy System." *Bul. Atomic Sci,* November, 1978.

Eppler, E. P. *Nuclear Safety* 11, 323, 1970.

Federal Power Commission. "Underground Storage of Natural Gas by Interstate Pipeline Companies," Washington, D.C.: U.S. Federal Power Commission, 1971, as quoted in Penner, 1975, p. 179.

Feiveson, H. A.; Taylor, T. B.; Von Hippel, F.; and Williams, R. H. "The Plutonium Economy, Why We Should and Can Wait." *Bul. Atomic Sci.* 32, 10, December, 1976, p. 10.

Fenner, D. "Power from the Earth." *Environment* 13, 19, December, 1971.

Flowers, B. "A Watchdog's View: Nuclear Power and the Public Interest." *Bul. Atomic Sci.* 32, 10, December, 1976, p. 22.

Ford Foundation. *Energy Policy Project, A Time to Choose.* Cambridge, Mass.: Ballinger, 1974.

Forsythe, E. Brookhaven National Laboratory, personal communication, 1977.

Fowler, J. M. *Energy and the Environment.* New York: McGraw-Hill, 1975.

Fultz, D.; Long, R. R.; Owens, G. V.; Bohan, W.; Kaylor, R.; and Weil, J. *Studies of Thermal Convection in a Rotating Cylinder with Some Implications for Large-scale Atmospheric Motions.* Meteorological Monographs, Amer. Meteorological Society, 1959.

Golding, E. W. *The Generation of Electricity by Wind Power.* New York: Philosophical Library, 1956. Reprint, London: Spon, Ltd.; New York: Wiley and Sons, 1976.

> An important source. The reprint has a little additional material on wind structure and lists large wind turbines of the period 1956 to 1965.

Goody, R. M., and Walker, J. C. G. *Atmospheres.* Englewood Cliffs, N.J.: Prentice-Hall, 1972.

Gray, T. J., and Gashus, O. K., eds. *Tidal Power.* New York: Plenum Press, 1972.

> Especially, articles by F. L. Lawton and T. L. Shaw.

Gregory, D. P. *A Hydrogen-Energy System.* Arlington, Virginia: American Gas Association, 1973, as quoted in Penner, 1975, p. 246.

> At Ludington pumped storage facility, capital cost storage installation was more than $6,100/10^6$ Btu$_e$, or, including associated transmission network, it was $202/kw$_e$. These correspond to total cost $330 and $380 million, respectively.

Gruener, G. "Subsidizing Solar." *Bul. Atomic Sci.* 34, 1, January, 1978, p. 2.

> Exporting nuclear reactors decreases our national security. Subsidizing export of solar-related energy sources to Third World coun-

tries to undercut foreign nuclear export has several national-security as well as domestic advantages.

Hallett, N. C. "Study, Cost and System Analysis of Liquid Hydrogen Production." Air Products and Chemicals, Inc., NASA Contractor Report CR-73226, Washington, D.C., 1968, as quoted in Penner, 1975, p. 179.

Hammond, A. L. "Nuclear Moratorium: Study claims that effects would be modest, foresees low growth rate of total energy demand." *Science* 195, 156, January 14, 1977a.

———. "Soft Technology Energy Debate: Limits to Growth Revisited?" *Science* 196, 961, May 27, 1977b.

———. "Photovoltaics: The Semiconductor Revolution Comes to Solar." *Science* 197, 445, July 29, 1977c.

———, and Metz, W. D. "Solar Energy Research: Making Solar After the Nuclear Model?" *Science* 197, 241, July 15, 1977d.

   NASA-NSF-ERDA solar program started largely to find practical spin-off from space program. Authors favor more emphasis on local on-site solar, less on centralized plants such as power tower. Wind receives only 8 percent of solar budget. ERDA leadership and OMB ruled out parallel development of competing concepts within each program. Some charge that this means solar is set up to fail. ERDA people consider wind competitive at current electric rates (or at least within a factor 2), other solar-related sources within factors 2 to 40.

———. "An International Partnership for Solar Power." *Science* 197, 623, August 19, 1977e.

   In underdeveloped countries beyond large power grids shaft power is needed but electricity costs ten to a hundred times as much as in United States. Huge market for cost-effective use of wind and other solar power. International cooperation needed to accomplish it.

———. "Photosynthetic Solar Energy: Rediscovering Biomass Fuels." *Science* 197, 745, August 19, 1977f.

   About 3 10^8 tons of agricultural wastes are "potentially collectable" per year, enough to generate about 4 quads (or 2 percent of our total energy use). Hydrogen could be produced economically from organic wastes by use of steam from a solar-heated boiler.

———. "An Interim Look at Energy." *Science* 199, 607, February 10, 1978.

   Introduction to a wide-ranging sequence of eight articles on energy sources to meet the world need. One of them, "Water and Energy," mentions wind power favorably as not requiring cooling water. Otherwise the only mention of wind power is on a 10-kilowatt scale for rural communities in the Third World—one more indication of the persistent lack of appreciation of large-scale wind power.

Harwood, S., et al. "The Cost of Turning It Off." *Environment* 18, 10, December, 1976, p. 17.

Hawkes, N. "Benn and British Rethink Energy Policy." *Science* 196, 146, April 8, 1977.

Britain has 40 percent more electric capacity than it needs, so is now going slow on nuclear. Investigating wave power, for which it is well situated, in a small way.

Hayes, D. *Rays of Hope.* New York: W. W. Norton, 1977.
Tvind, a Danish College, has nearly completed a 2-megawatt wind turbine at an artificially low cost of $350,000.

Herman, S. W., and Cannon, J. C. *Energy Futures—Industry and the New Technologies.* New York: INFORM, INC., 1976.
On wind power, describes plans and cost estimates of four large firms, that frankly await government financing, and one small one (Zephyr Wind Dynamo Co., Box 241, Brunswick, Maine).

Heronemus, W. E. "Pollution Free Energy from Offshore Winds." Reprint, 8th Annual Conference and Exposition, Marine Technology Station, Washington, D.C., 1972.

———. "Wind Power: A Near-Term Partial Solution to the Energy Crisis." In Ruedisili, 1975, p. 375.

Hewson, E. W. "Generation of Power from the Wind." *Bul. Am. Meteorolog. Soc.* 56, 7, 660, July, 1975.
Takes the total power of the wind available to man as either 10^7Mw (by assuming it an arbitrary factor 10^7 less than the atmosphere's total) or, following Von Arx (1974), 10^6Mw. Proposes thousands of wind dynamos where few people will see them. With mass production commencing in 1982 following present federal demonstration program, it should be possible in the 1990s to produce 10 percent of United States energy requirements by WECS. Proposes a "wing generator" in which a stack of wings aloft, each with aspirator holes drawing air by Bernoulli's principle, replaces the rotating hollow blades of figure 11—a high-solidity and expensive way to intercept wind.

Hide, R. *Quart. J. Roy. Meteorology Soc.* 79, 161, 1953.

Hildebrandt, A. F., and Vant-Hull, L. L. "Power with Heliostats." *Science* 197, 1139, September 16, 1977.
Estimates future cost of power tower installations as $1700 per installed kilowatt with six-hour storage when production of installations reaches 800 Mw/year. Capacity factor at least 41 percent would be competitive with nuclear power having 61 percent capacity factor, "U.S. regulations essentially forbid the utilities to invest in new technology until it is proven over a period of time," hence need for government promotion.

Hill, D. S. "Reaping the Wind." *Washington Post,* Outlook section, September 12, 1976, p. C1.

Hinrichsen, D., and Cawood, P. "Fresh Breeze for Denmark's Windmills." *New Scientist,* June 10, 1976.

Hirst, E. "Residential Energy Use Alternatives: 1976 to 2000." *Science* 194, 1247, December 17, 1976.
Oak Ridge computerized study finds annual energy use growth rate as 2.5 percent in business-as-usual forecast with no conservation,

0.4 percent as low forecast with reasonable conservation but no household solar or change of life style. Most of reduction comes from improved household equipment.

Hogerton, J. F. "The Arrival of Nuclear Power." *Sci. Amer.* 218, February, 1968, p. 21.

Hohenemser, K. H. "Wind Power Update." *Environment* 19, 1, January 5, 1977.

Holdren, John P. "Risk of Renewable Energy Sources: A Critique of the Inhaber Report." Berkeley, Cal.: Energy Resources Group, 1979.

Hubbert, M. K. "Energy Resources of the Earth." *Sci. Amer.* 224, September, 1971, p. 62.

Hütter, U. "Optimum Design Concept for Windelectric Converters." Workshop on Advanced Wind Energy Systems, Stockholm, August 29, 1974.

————. *Vom Wert der Windenergie, Forschungsinstitut Windenergietechnik.* University of Stuttgart, 1975.

Inglis, D. R. *Rev. Mod. Phys.* 27, 212, 1955; *Journal of Geomagnetism and Geoelectricity* 17, 517, 1965.

————. "Nuclear Energy and the Malthusian Dilemma." *Bul. Atomic Sci.,* February, 1971, p. 15, and June, 1971, p. 39.

————. *Nuclear Energy: Its Physics and Its Social Challenge.* Reading, Mass.: Addison-Wesley, 1973.

————. "Power from the Ocean Winds." *Environment* 20, October, 1978.

————. "Wind Power Now." *Bul. Atomic Sci.* 32, 8, 1975, p. 21; *The Rochester Engineer* 54, 10, May, 1976, p. 182.

————. "Nuclear Energy: Rasmussen Reviewed." *Environment* 18, November, 1976, p. 38.

————. "An Answer is Blowing in the Wind." *The Progressive* 41, 1, January, 1976, p. 43.

————. "Community Energy System." *Bul. Atomic Sci.,* April, 1979.

Inglis, D. R., and Ringo, G. R. "Underground Construction of Nuclear Reactors," Argonne National Laboratory Report ANL-5652, 1957.

Inhaber, H. "Risk of Energy Production." Atomic Energy Control Board of Canada Report AECB-1119, 1978 (with addendum on wind power); reprinted in *Science* 203, 1979.

JBF Scientific Corp. "Summary of Current Cost Estimates of Large Wind Energy Systems." ERDA Report DSE/2521-1, April, 1977a.

Illustrates ERDA emphasis on immediate commercial economics. Estimates that, in 8-meter/second wind regime, megawatt-scale wind power added to commercial grids just saves money now from fuel oil savings alone, with no credit for capital cost replacement. Saves more later as oil prices rise. G.E. study estimated cost of megawatt-scale wind dynamo as $405/kw (1975 dollars, plus $27/kw for installation) with production run of 100 units. Aerospace company estimates higher.

————, ed. "Third Wind Energy Workshop," Washington, D.C., September 1977. DOE report CONF 770921, 1977b.

Johnson, L. "Wise-Wind." *Wind Power Digest* 9, 24, Summer, 1977.

Kahn, H.; Brown, W.; and Martel, L. *The Next 200 Years.* New York: William Morrow, 1976.

Kendall, H. W., et al. "The Risks of Nuclear Power Reactors, a Review of 'The Reactor Safety Study WASH-1400.' " Cambridge, Mass.: Union of Concerned Scientists, 1977.

Kilar, L. A. "Design Study and Economic Assessment of Multi-Unit Offshore Wind Energy Conversion Systems Application." Westinghouse Electric Corporation Report WASH/2230-78/4, UC-60, June, 1979.

Kottler, R. J., ed. *Proceedings of the Fourth Biennial Workshop of Wind Energy Conversion Systems, CONF-791097.* Washington, D.C.: United States Department of Energy, 1979.

Kuwada, J., and Ramey, H., Jr. "The Challenge of Geothermal Energy, in Energy, Environment, Productivity." In *RANN Symposium Proceedings,* edited by J. Holmes, Washington: National Science Foundation, 1973.

Lanou, R. "Nuclear Plants: The More They Build, the More You Pay." Washington, D.C.: Center for Responsive Law.
> Recent reactor capital estimates (table 3–7) for completion dates 1978 to 1985 range from about $900 to $1300 per installed kilowatt. Other aspects of nuclear and coal power costs are discussed.

Lapp, R. E. *A Citizen's Guide to Nuclear Power.* Washington, D.C.: The New Republic, 1971.

Lapuy, L. "Solar Energy Planning Being Slighted." *Wall Street Journal,* February 26, 1976.

"Lockheed Study Proposes WECS Fertilizer Production Units." *A.W.A.E. Newsletter, American Wind Energy Association,* Spring, 1977, p. 12.
> A 2-megawatt wind dynamo could produce enough ammonium nitrate, that is now produced from natural gas, to fertilize 23,000 acres for one crop year.

London, J., and Sasamori, T. "Energy Budget of the Atmosphere." In Matthews, 1971, 141.

Lorenz, E. N. "The Nature and Theory of the General Circulation of the Atmosphere." World Meteorological Organization, 1967.

——. "Climatic Changes as a Mathematical Problem." In Matthews, 1971, p. 179.

Lovins, A. B. *World Energy Strategies.* Cambridge, Mass.: Ballinger, 1975a.

——. "Energy Strategy: The Road Not Taken?" *Foreign Affairs,* October, 1976, p. 65. See also author's discussions with critics: *Foreign Affairs,* April, 1977, and "Solar Numbers and Hard Assumptions," *Not Man Apart* 7, 11, June, 1977, p. 4.

——, and Price, J. H. *Non-nuclear Futures.* New York: Ballinger (and Friends of the Earth), 1975b.

——. *Soft Energy Paths: Toward a Durable Peace.* Cambridge, Mass.: Ballinger (and Friends of the Earth), 1977.

McCaull, J. "Windmills." *Environment* 15, 1, January, 1973, p. 6.
——. "Energy and Jobs." *Environment* 18, 1, January 18, 1976.
McCrawley, J. "Dynaship." *Rudder,* November, 1971.
McGowan, J. S. "Ocean Thermal Energy Conversion, A Significant Resource." *Solar Energy* 18, 1976.
Matthews, W. H.; Kellogg, W. W.; and Robinson, G. C., eds. *Man's Impact on Climate.* Cambridge, Mass.: MIT Press, 1971.
Meadows, D. H., and Meadows, D. L. *Limits to Growth.* Potomac Associates Book, Universe, 1972.
Meinel, A. B., and Meinel, M. P. *Bul. Atomic Sci.* 27, 7, October, 1971, p. 32.
——. "Energy Transfer in a Large-Scale Thermal Solar Power Farm." *Solar Energy* 18, 177, 1976.
——. "Physics Looks at Solar Energy." *Physics Today,* February, 1972, p. 44.
Merriam, M. F. "Wind Energy for Human Needs." *Technology Review,* January, 1977.
Metz, W. D. "Solar Politics." *Science* 194, 1256, December 17, 1976.
——. "Reprocessing Alternatives: The Options Multiply." *Science* 196, 284, April 15, 1977a.
——. "Wind Energy: Large and Small Systems Competing." *Science* 197, 973, September 2, 1977b.
——. "Ocean Thermal Energy: The Biggest Gamble in Solar Power." *Science* 198, 180, October 14, 1977c.
——. "Solar Thermal Electricity: Power Tower Dominates Research." *Science* 197, 353, July 22, 1977d.
Meyer, H. "Synchronous Inversion." *Windworks,* Mukwonago, Wisconsin, February 9, 1976.
Miller, O. A. "Windchargers You Can Buy Now." *Mechanix Illus.* 72, 582, November, 1976, p. 31.
Morgan, R. "Nuclear Power: The Bargain We Can't Afford." Environmental Action Foundation, 1977.
Muffler, L. J. P., and White, D. E. "Geothermal Energy." *The Science Teacher* 39, 1972, p. 3.
Musgrove, P. J. "The Variable Geometry Windmill." *Wind Power Digest* 8, 27, Spring, 1977.
Myers, D., III. *The Nuclear Debate.* New York: Praeger, 1977.
Nelson, V., and Gilmore, E. "Potential of Wind Generated Power in Texas." Project N/T-8, Governor's Energy Advisory Council, State of Texas, October 15, 1974.
> An early exercise in thinking big, Texas style, that has not yet caught on. Projects 250 thousand megawatts average from the Panhandle. Includes wind surveys.
Newark, N.J. Electric Power Research Institute. "Newark, An Assessment of Energy Storage Systems Suitable for Use by Electric Utilities." Prepared by Public Service Electric and Gas Company, Newark, N.J. EPRI-EM-264 and ERDA E(11-1)-2501, July, 1976.

Newman, B. G. "The Spacing of Wind Turbines in Large Arrays." Montreal: Dept. of Mechanical Engineering, McGill University. *Energy Conversion, an International Journal,* vol. 16, pp. 169–71, London: Pergamon Press, 1977.

Novick, S. *The Electric War: The Fight over Nuclear Power.* Sierra Club Books, 1977.

Odum, H. T., and Odum, E. C. *Energy Basis for Man and Nature.* New York: McGraw-Hill, 1976.

OTA, "Application of Solar Technology to Today's Energy Needs." Office of Technology Assessment, U.S. Congress, Washington, D.C., 1977.

Papagianakis, S. "Aeolian Energy in Greece." In *Solar and Aeolian Energy* (Proceedings of an international seminar at Sounion, Greece, September, 1961), edited by A. G. Spinides. New York: Plenum Press (distributor).

Penner, S. S., and Icerman, L. *Energy.* Vol. 1, *Demands, Resources, Impact, Technology,* 1974. Vol. 2, *Non-nuclear Energy Technologies,* 1975. Vol. 3, *Nuclear Energy and Energy Policies,* 1976. Reading, Mass.: Addison-Wesley.

Petruschell, R. L., and Salter, R. G. "Electricity Cost Model for Comparison of California Power Plant Siting Alternatives." Santa Monica, RAND, R-1087-RF/CSA, January, 1973.

Pinson, "Cycloturbine." *Wind Power Digest,* Fall, 1977, p. C14.

Post R., and Post, S. "Flywheels." *Sci. Amer.* 226, 6, 1973 p. 17.

"Pumped Storage to the Rescue." *Engineering News Record,* June 3, 1971, p. 16.

Putnam, P. C. *Power from the Wind.* New York: Van Nostrand, 1948.

RANN, "Solar Energy Research Program Alternatives, Research Applied to National Needs Program." National Science Foundation NSF/RA/N-73-111B, December, 1973.

Ray, D. L. "The Nation's Energy Future." Report WASH-1281, U.S. Atomic Energy Commission, 1973.

Reed, J. W. "Wind Climatology." In Eldridge, 1975, pp. 319 and 325 and fig. 3; "Wind Power Climatology in the U.S." Sandia Laboratories SAND 74-0348, June, 1975.

Research and Education Association. *Modern Energy Technology.* Vol. 2. New York: REA, 1976.

Rex, R. W. "Geothermal Energy—Neglected Energy Option." *Bul. Atomic Sci.* 27, 8, October, 1971, p. 52.
    Estimates geothermal energy potential of United States as 10^5 to 10^6Mw within thirty years.

Rockefeller Fund, California State Assembly, RF/CSA, January, 1973.
    Power transmission line for 1250 Mva at 500 kv with steel towers installed in typical California terrain costs $150,000 per mile (1973 dollars).

Rogers, F. C. "Underground Nuclear Power Plants." *Bul. Atomic Sci.,* October, 1971, p. 38.

Rose, D. J. "Nuclear Electric Power." *Science* 184, 357, April 19, 1974.

Ruedisili, L. C., and Firebaugh, M. W., eds. *Perspectives on Energy.* New York: Oxford Press, 1975.

Ryle, Sir M. "The Economics of Alternative Energy Sources." *Nature,* May 12, 1977.
> Compares economics of wind and nuclear power under British conditions, concluding, as in this book for United States, that large-scale wind power is ready and urgently needed. Use of storage to meet peaking needs reduces cost of nuclear power by 50 percent. Storage being included, cost of wind power half that of nuclear. Above-ground sites for pumped storage being insufficient (underground not considered) and much of energy being used for heating, local thermal storage is favored. Synchronizing output of variable speed alternators by use of modern solid state circuitry reduces cost of future wind dynamos. Quick deployment of wind power on British west coast and in North Sea shallows needed to save North Sea oil for future petrochemicals.

Salter, S. H. "Wave Power." *Nature* 249, 720, 1974.

Schelling, T. C. "The Promise and the Curse." *Saturday Review,* January 22, 1977, p. 26.
> Claims the bargaining strength of the United States government in coping with the worst evils of proliferation depends on our position as a major supplier.

Scott, F. M. "Underground Hydroelectric Pumped Storage: A Practical Option." *Energy,* Fall, 1977, p. 20.

Seams, R. C. Jr. "Energy and Marine Technology." ERDA *Weekly Announcements* 2, 49, December 17, 1976.

Shapley, D. "Reactor Safety: Independence of Rasmussen Study Doubted." *Science* 197, 29, July 1, 1977.

Sheperdson, W. "W.T.G. Energy Systems: 200 kw for Cuttyhunk Island." *Wind Power Digest,* Fall, 1977, p. 6.

Smithsonian Institution, *Annual Report,* 1913.

Sørensen, B. "Wind Energy." *Bul. Atomic Sci.* 32, 8, September, 1976a: "Dependability of Wind Energy Generators with Short-Term Energy Storage." *Science* 194, 935, November 26, 1976b.

Steinhart, J. S., et al. "A Low Energy Scenario for the United States, 1975–2050." IES report 83. Madison: Institute for Environmental Studies, University of Wisconsin, 1977.

Templin, R. J. "An Estimate of the Interaction of Windmills in Widespread Arrays." LTR-LA-160, National Research Council of Canada, June, 1974.

Thirring, H. *Windmills to Nuclear Power.* New York: Greenwood Press, 1968.
> This suggests as an alternative title for the present volume, *Nuclear Power to Windmills.*

Thomas, P. H. "The Wind Power Generator, Twin Wheel Type." Federal Power Commission, 1946.

Vendvyes, L. A. "Superphenix, A Full-Scale Breeder Reactor." *Sci. Amer.* 236, 3, March, 1977, p. 26.

von Arx, W. S. "Energy: Natural Limits and Abundances." *Transactions of the American Meteorological Society* 55, 828, October, 1974.

von Hippel, F., and Williams, R. H. "Toward a Solar Civilization." *Bul. Atomic Sci.* 33, 8, October, 1977, p. 12.

Wahrenbrock, H. E. "Can We Afford *Not* to Develop the High-Tower Windmill Now?" *Public Utilities Fortnightly,* September 17, 1979.

Wallace, D., ed. "Riding the Wind." In *Energy We Can Live With.* Emmaus, Pa.: Rodale Press, 1976, p. 22.

Warshay, M., and Wright, L. O. "Cost and Size Estimates for an Electrochemical Bulk Energy Storage Concept." NASA TM X-3192, February, 1975.

Wasserman, H. "Nuclear Energy at the Grass Roots." *The Nation,* February 26, 1977, p. 245; "Industry Hides from the Sun." *The Nation,* March 5, 1977, p. 263.

Watson, M. B., et al., "Underground Nuclear Power Plant Siting." EQL report no. 6, Environmental Quality Lab, California Institute of Technology, 1972.

Weatherholt, L., ed. "Vertical-Axis Wind Turbine Technology Workshop." Sandia Laboratories Report SAND 76-5586, 1976.

Webb, R. E. *The Accident Hazards of Nuclear Power Plants.* Amherst: University of Massachusetts Press, 1975.

Weinberg, A. M. "The Maturity and Future of Nuclear Energy." *American Scientist* 64, January–February, 1976, p. 16.

> "Being an optimist, I hope that by the year 200 the world will have a choice of six breeders and near-breeders. I confess to being less optimistic about fusion (not to speak of geothermal and solar energy), but this is undoubtedly a prejudice that can be expected, if not excused, in one who has devoted his career to the development of fission. . . . To reject nuclear energy while invoking such magical talismans as fusion, or solar or geothermal energy, seems unwarranted."

———. "Nuclear Energy and the Environment." *Bul. Atomic Sci.,* June, 1970, p. 69.

Welch, B. L. "Fission: The Faustian Dream. Nuclear Energy on the Dole." *The Nation,* February 26, 1977, p. 231.

Wellikoff, A. "In Sunspot." *Not Man Apart* 7, 12, June, 1977, p. 14.

Westh, H. C. "A Comparison of Wind Turbine Generators." In Eldridge, 1975, p. 156.

Wilrich, M., and Taylor, T. B. *Nuclear Theft: Risks and Safeguards.* Cambridge, Mass.: Ballinger, 1974.

Wilson, R. E., and Lissaman, P. B. S. "Applied Aerodynamics of Wind Power Machines." Oregon State University, 1974.

Wilson, R., and Jones, W. J. *Energy, Ecology, and the Environment.* New York: Academic Press, 1974.

Wood, L. "Fusion Power." *Environment* 14, 29, May, 1972.
Zelby, L. W. "Don't Get Swept Away by Wind Power Hopes." *Bul. Atomic Sci.* 32, 3, March, 1976, p. 59.
> Criticizes ref. (Inglis, 1975) for using large-production-run cost estimates. Assumes only storage method is hydrogen. Finds limitation in the claim made by E. W. Hewson, in "Generation of Power from the Wind" (1975), that total available wind energy is only 10 percent of year-2000 electric energy consumption. This book's appendix 1 and chapter 4 may be considered as reply.

Zener, C. "Solar Power from Seawater." *Physics Today,* January, 1972.

Index